CRYSTALS, ELECTRONS, TRANSISTORS

FROM SCHOLAR'S STUDY TO INDUSTRIAL RESEARCH

Michael Eckert
Helmut Schubert

Translated by Thomas Hughes

AIP
American Institute of Physics New York

Originally published as *Kristalle, Elektronen, Transistoren*

Library of Congress Cataloging-in-Publication Data

Eckert, Michael
Crystals, electrons, transistors.

Translation of: Kristalle, Elektronen, Transistoren.
Bibliography: p.
Includes index.
1. Science–History. 2. Physics–Research–History. 3. Research, Industrial–History. I. Schubert, Helmut. II. Title.
Q125.E36313 1989 530′.09 89-14941
ISBN 0-88318-622-5
ISBN 0-88318-719-1 (pbk.)

CONTENTS

PREFACE TO THE AMERICAN EDITION

READERS SHOULD BE AWARE that this book was first written for the German readership. Our intention was to present a more "down to earth" history of science than it is often accounted for with the histories of great geniuses and their achievements; at the same time, we wanted to attract the German readers' attention to relevant developments in the USA—such as the U.S. radar project during World War II and industrial research at the Bell Laboratories—which were not yet reported in the German literature. American readers who are familiar with books such as Daniel Kevles' *The Physicists: The History of a Scientific Community in Modern America* or with the recent volumes of *Historical Studies in the Physical and Biological Sciences*, therefore, will perhaps find Chapters 4 and 5 less interesting and focus their attention to other chapters, where we describe the German situation in more detail.

Science and scientists are subject to social, economic, and political influences. What seems evident in today's scientific actualities (e.g., the military impact in cases like SDI or nuclear affairs) is the result of a historical process, where national characteristics, international ramifications, institutional and personal features as well as other social and intellectual factors are entangled into a pattern of many facets. In view of its enormous complexity, any attempt to describe it by way of simple lessons is doomed to failure. Nevertheless, it is possible to portray some of these facets and get pertinent impressions.

Our method of accounting for scientific development, therefore, was to illuminate some facets in extreme detail—by quoting from original source material and publications—and thereby generate glimpses of historical events with as much authenticity as possible. When the degree of detail seems exaggerated in some instances, this should be considered as our effort to transmit, beyond the mere facts, a feeling for styles and attitudes without which science may not be duly described. For example, we hope that the American reader will enjoy what we have included in order to portray German professional performance.

Readers with a deeper interest in the intellectual history of solid state physics are referred to the forthcoming publication *Out of the Crystal Maze* (Oxford University Press).

INTRODUCTION

FOR CENTURIES CRYSTALS WERE considered to be the mysterious raw material from which the world was made. Aristotle described the universe as a gigantic sphere made up of concentric crystal shells. Their purity and perfection set crystals apart from other solid materials, and crystals still are objects of fascination for naturalists and collectors because of their transparency and their wealth of colors and shapes (Fig. 1).

Crystals lost some of their magic at thc turn of the century, when we learned more about the structure of matter, about atoms and electrons, and began to use x rays, electron beams, and neutrons to study crystal structure.

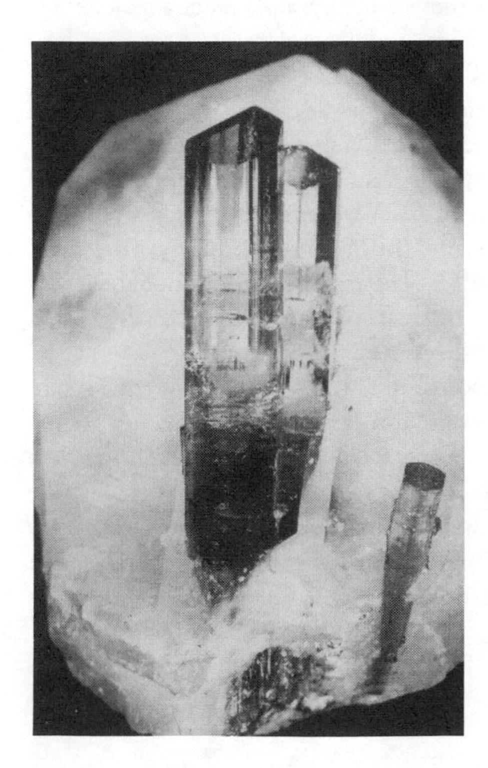

FIG. 1. A large full-color picture of this reddish tourmaline imbedded in white quartz adorned a 1984 calendar of the WELEDA anthroposophical drug firm. Tourmaline is outstanding among the minerals of the earth for the rich diversity of its colors and the wide range of its component materials. In anthroposophy special healing powers are attributed to drugs made from minerals.

Today artificial crystals are "grown" in laboratories—not because of their fascinating shapes and colors but because they are used in making numerous kinds of electronic components. Transistors, integrated circuits, chips, light diodes, and many other products of modern technology come from this branch of industry, which grew up practically overnight and is still growing. The stormy development of this industry has been based in large measure on a branch of science called solid-state physics, which for about 50 years has been concerned with the study of crystals and electrons.

The close link of this field of physics to practical applications makes it an excellent candidate for a case study of scientific development as influenced by technology. Although the intellectual heritage of physics extends back to ancient times, its broad societal role as an independent profession with institutionally guaranteed educational and career opportunities is a very recent situation. With a few individual early exceptions, industrial physicists have been around only since the years following World War I, when discoveries in physics opened up technical opportunities for new products and production processes. Investigations by industrial physicists are responsible for the use of new materials, such as light metal alloys in airplane construction and special conductor and insulator materials in the electrical industry. Industrial physicists made a major effort to study the magnetic behavior of transformer plates and to search for long-lived filaments for light bulbs. The physics profession has developed an important body of knowledge on which large segments of modern industry are based, and which has been incorporated into the curricula of professional schools.

As late as about 150 years ago the progress of natural science often depended on feudal patronage, as in the case of Josef Fraunhofer,the glass polisher, optician, instrument maker, and member of the Bavarian Academy of Sciences. Today government and industry provide for the support of professional scientific research, research which serves the interests of state and the economy in a manner which clearly earns the support received.

This book does not describe the spectacular events of the history of physics: neither Einstein's stroke of genius in formulating his theory of relativity nor the discovery of the fission of the atomic nucleus by Otto Hahn are discussed. The discussions that follow are concerned much more with events that demonstrate the growth of physics into a profession and the increasing industrial importance of this branch of science, especially in the area of solid-state research. The character of this science, which has become "big science" (Derek de Solla Price), is not reflected only in big things like nuclear weapons, atomic reactors, and elementary particle accelerators. The tendency to emphasize these things gives the wrong appearance of normal day-to-day physical research and gives physicists a false image in the minds of many people. According to estimates from the West German Institute for Labor-Market and Vocational Research, about 10% of the 20,100 physicists registered in the Federal Republic of Germany in 1980 (19,700 of them males) were designated as "solid-state physicists." When these are lumped with the related disciplines of "electronics" and "data processing," they account for almost one-third of all physicists (29.1%), as compared with only 11.4% in the "big-science" fields of "nuclear physics," "reactor physics/reactor construction," and "high-energy physics." Solid-state physicists are involved in many fields and produce a wide variety of products, such as transistors and the more miniaturized semiconductor components (integrated circuits and chips). These make it possible to increase efficiency, automate, and computerize many everyday activities.

The concept of "solid-state physics," however, has remained almost unknown to the public, even though the results of this science have had effects on the life of every individual person. This is undoubtedly evidence of a lack of understanding of the historical roots of modern technologies. This lack is particularly unfortunate in training institutions, where microelectronics are part of the everyday routine. This book is intended to help remedy that lack.

An extensive history of solid-state physics is in preparation at the present time.[92] The present book is based on work that was done within the scope of an international project on the history of solid-state physics. The work benefitted from the support of the Volkswagen Foundation and access to unpublished source material for the project "History of Solid State Physics" (Deutsches Museum, Munich) and manuscripts written by colleagues on the project that will be published shortly (Henriksen, Hoch, Keith, and Teichmann). Further thanks are due to David Cassidy, Günther Küppers, and Paul Forman, who were not connected with the project but who read the manuscript and provided professional criticism.

CHRONOLOGY

Date	Developments in the field of solid-state physics, electronics, and industrial research	Date	General historical, sociological, and technological events
AD		AD	
		1657	Academia del Cimento (Florence); beginning of the establishment of scientific academies with goals oriented toward the practical sphere.
		1662	Royal Society (London)
		1666	Académie des Sciences (Paris)
1729	"Elementa physicae" (Musschenbroek); flowering of instrumentation and experimental physics in Holland.		
		1735	Schemnitz Mining Academy; beginning of the establishment of specialized scientific and technical schools.
1745	Experiments with the "Leyden jar" (Musschenbroek).		
		1747	École des Ponts et Chaussées.
1780	Experiments on "Animal electricity" with frog legs (Galvani).		
		1782	Development of the reciprocating steam engine (Watt).
		1783	École des Mines.

Date	Developments in the field of solid-state physics, electronics, and industrial research	Date	General historical, sociological, and technological events
		1794	École Polytechnique; model for the institutes of technology established in the 19th century.
1795	First electric battery (Volta).		
1819	Lectures at the École des Ponts et Chaussees on the strength of bridges and elasticity theory (Navier).		
		1822	Founding of the Gesellschaft Deutscher Naturforscher und Ärtzte (GDNÄ) (Society of German Scientific Investigators and Physicians), the first scientific lobbying organization.
1824	Theory on the conversion of heat into mechanical energy (Carnot).		
		1825	Establishment of the first German institute of technology in Karlsruhe.
1825	Discovery of "Ohm's law."		
		1831	Founding of the British Association for the Advancement of Science (BAAS), patterned after the GDNÄ.
		1847	Pointer telegraph (W. Siemens).
		1851	First world exhibition in London.
		1854	First use of arc lamps for nighttime construction work in Paris.
1859	Discovery of spectrum analysis (Bunsen, Kirchhoff).		

Date	Developments in the field of solid-state physics, electronics, and industrial research	Date	General historical, sociological, and technological events
		1861	Bessemer process for optimizing steel production.
		1866	First operating telegraph connection between the continents of Europe and America.
1869	Discovery of cathode rays (Hittorf).		
		1887	Establishment of the Imperial Institute of Physics and Technology (Physikalisch-Technische Reichsanstalt, PTR), beginning of state support of research for industry. Model for the National Bureau of Standards (United States), the National Physical Laboratory (Great Britain), and Riken (Japan).
1895	Discovery of x rays (Röntgen).		
1897	Discovery of radio-activity (Becquerel). Discovery of the electron as the particle in cathode rays (J. J. Thomson).		
		1898	Establishment of the Göttingen Association (Göttinger Vereinigung), first example of industry support of research at universities.
About 1900	Studies of radiant heat at the Physikalisch-Technische Reichsanstalt (Pringsheim, Lummer). Discovery of Planck's radiation law. First (classical) electron theory of metals (Drude; J. J. Thomson).		

Date	Developments in the field of solid-state physics, electronics, and industrial research	Date	General historical, sociological, and technological events
		1900	Tenth world exhibition and first international congress of physicists in Paris.
1902	Establishment of the Research Laboratory of the General Electric Company.		
1907	Consolidation of Bell's research into a central laboratory.		
1908	Liquefaction of helium at − 269 °C (Kammerlingh-Onnes).		
1909	Invention of the light metal alloy "Duralumin" (Wilm).	1909	Founding of "Zeppelin, GmbH" for building airships.
1911	Discovery of superconductivity (Kammerlingh-Onnes). First Solvay Congress ("official" prelude to the quantum theory).		
		1911	Founding of the Kaiser-Wilhelm Gesellschaft in Berlin.
1912	Discovery of x-ray diffraction in crystals ("Laue experiment").		
1913	The Bohr atom.		
1914	The Siemens firm builds a central research laboratory.	1914	First World War.
		1915	Department of Scientific and Industrial Research (Great Britain).
		1916	National Research Council (United States).
1919	"Atomic structure and spectral lines" (Sommerfeld).		
		1920	Founding of the German Science Aid Society (Notgemeinschaft der

Date	Developments in the field of solid-state physics, electronics, and industrial research	Date	General historical, sociological, and technological events
			Deutschen Wissenschaft) and the Helmholtz Society Helmholtzgesellschaft).
		Since 1924	Support of international scientific exchanges in the United States by the International Education Board of the Rockefeller Foundation. Expansion in the American university community; United States becomes a new great power in science.
1925	Quantum mechanics (Heisenberg, Schrödinger).		
1927	Demonstration of electron deflection in crystals (Davisson and Germer at Bell Laboratories).		
Since 1927	Development of a quantum-mechanical electron theory of metals, semiconductors, and insulators.		
		1929	Economic crash.
Since 1930	Theories on insulator crystals (Bristol University). Pohl's "color center" research (Göttingen University).		
		1933	Nazis seize power; April 7: Law on "Restoration of the Career Civil Service" forces emigration of Jewish and leftist scientists.

Date	Developments in the field of solid-state physics, electronics, and industrial research	Date	General historical, sociological, and technological events
1938	Demonstration of a three-electrode crystal (Pohl, Hilsch).		
1939	Boundary layer theory to explain the semiconductor-metal contact effect (Schottky).	1939	Second World War.
1940	Founding of the MIT Radlab to coordinate Allied radar research.		
		1941	Central direction of research of importance to the war effort in the United States by the Office of Scientific Research and Development (OSRD).
1942	Development of semiconductor diodes as radar detectors (Bell, Purdue).		
		1944	Launching of V-2 rockets against Paris and London.
1945	Completion of the first electronic digital computer for ballistic calculations (ENIAC) at MIT.	1945	Atomic bombs dropped on Hiroshima and Nagasaki.
1947	Invention of the germanium point-contact transistor at Bell Labs (Bardeen, Brattain, Schockley).	1947	Rocket plane reaches the speed of sound for the first time.
1948	Invention of the germanium junction transistor (Schockley).		
1950	U.S. Navy's "Project Tinkertoy" to miniaturize electronic components with vacuum-tube technology.	1950	Korean War: Increased military research output in the United States.
1951	Discovery of the III–V semiconductor (Welker).		
1952	Bell Labs Symposium on Transistor Manufacturing.	1952	Detonation of the first hydrogen bomb.

Date	Developments in the field of solid-state physics, electronics, and industrial research	Date	General historical, sociological, and technological events
1953	Development of the no-contact zone-pull method of producing pure silicon (Siemens, Western Electric).		
1954	Development of the first silicon transistor at Texas Instruments.		
1956	Development of mesa transistors at Bell Labs. Second Bell Labs symposium.		
		1957	Sputnik shock: the first launch of a satellite by the Soviet Union caused a sudden intensification of American space research; miniaturization of electronics is a part of these efforts.
1958	Invention of the first integrated circuit at Texas Instruments (Kilby).	1958	American Explorer satellite.
1959	Development of the planar technology at Fairchild.		
1960	Invention of the MOS transistor.		
		1961	U.S. President John F. Kennedy decides on the Apollo moon-landing program. First successful test of the Minuteman intercontinental rocket. U.S.S.R. astronaut Gagarin becomes the first man in space.
		1963	First successful test of a Polaris A-3 rocket, launched from submarines.
		1964	With the intervention of the United States in Vietnam, this theater of war also becomes a testing ground for new weapons systems.

Date	Developments in the field of solid-state physics, electronics, and industrial research	Date	General historical, sociological, and technological events
1965	Digital Equipment, Inc., builds the PDP-8, the first minicomputer produced in large numbers and priced under $20,000.		
1967	Civilian use of integrated circuits equals that of the military for the first time.	1967	First space research program of the Ministry of Research in the Federal Republic of Germany which, among other things, made government resources available for the development of electronics.
		1969	U.S. Astronaut Neil Armstrong becomes the first man to set foot on the moon during Apollo II mission.
1970	Development of the microprocessor by the Intel firm (Hoff).		
		1972	"Strategic Arms Limitation Talks" (SALT) between the United States and the Soviet Union.
1974	Hewlett-Packard HP-65, the first programmable pocket calculator. Programs were written on small magnetic cards.		
1975	Several firms attain a packing density of 100,000 components on one chip.		
		1976	First test flights of "intelligent" cruise missiles, which locate their targets independently by means of advanced radar techniques using preprogrammed terrain information.

Date	Developments in the field of solid-state physics, electronics, and industrial research	Date	General historical, sociological, and technological events
1977	First personal computer developed by the Xerox Corporation.		
		1979	First launch of the European "Ariane" rocket.
1980	IBM produces superconductive switching elements made of niobium with ultrafast switching times (10^{-9} s).		
		1981	Military projects predominate over civilian projects in U.S. research and development; the military share climbs from $18 billion to more than $30 billion between 1981 and 1984, while during the same period the civilian share drops to less than $15 billion.
1983	Invention of the Quiteron, a superconductive circuit with three connections, which works like a transistor (IBM).	1983	U.S. President Ronald Reagan introduces the militarization of space with the "Strategic Defense Initiative" (SDI). Industrial research groups expect from this program enormous contracts in advanced areas of technology (lasers, electronics, etc.).
1985	For the first time 1 million components are integrated on a single chip.	1985	Conference of European Foreign Ministers and Ministers of Research about a European high-technology association (EUREKA).
1986	Two IBM physicists win the Nobel prize for a new method (scanning tunneling microscopy), which allows us		

Date	Developments in the field of solid-state physics, electronics, and industrial research	Date	General historical, sociological, and technological events
	to observe microelectronic layers at atomic resolution.		
1987	Two IBM physicists win the Nobel prize for discovering high-temperature superconductivity.		

1

PHYSICS BECOMES A PROFESSION

THE EMANCIPATION OF NATURAL SCIENCE IN THE 18TH AND 19TH CENTURIES

IN THE 18TH CENTURY the Dutchman Pieter van Musschenbroek (1692–1761) was investigating phenomena such as thermal expansion, phosphorescence, magnetism, and electricity—subjects which would characterize him today as a physicist, or more precisely, as a solid-state physicist. But in Musschenbroek's day physics was not yet an independent discipline. The career of "physicist" was to appear only in the 19th and 20th centuries. Musschenbroek's career may not have been untypical for many investigators of nature in the 18th century: he came from a well-to-do family, which had earned recognition in the area of instrument building. He studied medicine at the University of Leyden. He was a practicing physician and at various times was a professor of mathematics, philosophy, medicine, and astronomy. His experimental lectures at the University of Leyden drew students from all over Europe. The lecture texts, which were published in book form, were translated into many languages.

Musschenbroek's writings reflect the spirit of the "new science" of Galileo, Newton, Huygens, and other natural scientists and philosophers of the 17th century. In his *Elementa physicae* (1734) he emphasized the necessity of observation and experiment in order to learn about things and their properties. Above all, carefully selected instruments should be used and all possible external influences such as location or the weather situation should be taken into account. Only when experiments that are repeated many times give the same results, can the natural law which governs the behavior of the objects be discovered. Musschenbroek also valued the method of deduction with the aid of mathematics—provided a natural law identified in this fashion is further verified by experiment. This philosophy inspired the establishment of many amateur societies for experimental scientific research in Holland and elsewhere.[190]

ACADEMIES FOR APPLIED SCIENCES

The Dutch colonial power supported by a wealthy commercial middle class provided, even in the 17th century, conditions that were comparatively favorable for the spread of the new science. Instrument-building knowledge, which was very important for navigation, was highly developed there. The predominant Calvinism, with its ethic focused on the here and now and on practical benefits, set no clerical restrictions. "In order to do scientific work, I plan to go to Holland," said the Galileo student Andrea Sarti in Bertold Brecht's "The Life of Galileo," and the following words were said by the sly Galileo, who knew the monks were listening: "Unfortunately there are countries which have withdrawn from the protection of the Church."[18] The University of Leyden, where Musschenbroek worked for a long time, was a leading center of the new science.

A progressive climate also prevailed in Calvinist Scotland. In total contrast to the old English universities of Oxford and Cambridge, whose elite colleges of that day were described as "exclusive and somewhat eccentric clubs," the Scottish universities were free of religious restrictions and able to direct science to questions of technology. In England, where the citizenry in the 17th century had begun to strip power from the clergy and the nobility, merchants, smaller landowners, and successful professionals such as physicians and lawyers became supporters of science. In addition to the Royal Society, which was founded as early as 1662, and whose scientific research was financed by member contributions (the monarchy paid nothing!), there appeared private organizations such as the academies of dissenters and the Lunar Society in Birmingham, whose members met on the night of the full moon, always in a different place. The declared goal of the Royal Society was to "perfect our knowledge of the things of nature and to improve all useful arts, methods of production, mechanical processes, and discoveries by experiments (and not to become involved in theology, metaphysics, moral teaching, politics, grammar, rhetoric, or logic)" (from Ref. 10, p. 426)—entirely in line with the spirit of the new leading circles of the middle class. In the Lunar Society, to which belonged, for example, the steam engine builders James Watt (1736–1819) and Matthew Boulton (1728–1809), the combination of wealth with a spirit of enterprise, technology, and new science was obvious.[22]

In contrast to the situation in England, Scotland, and the Netherlands, the "ancient régime" of feudalism prevailed in France in the 18th centruy. Here, as in other absolutist states, the climate was wrong for the survival of new science. There was no free spirit of enterprise based on private initiatives of private societies and organizations opposing the clerical and feudal establishment. In these countries the support of science was the business of the state and supported the objectives of the prevailing mercantile economic order. This system was based on an active trade and commercial policy to increase the amount of money in the country, thereby increasing the financial power of the lord of the land and, in the last analysis, the military strength of "his" state, in order to bring gold into the country through the export of finished products. The mining industry was supported in order to maximize production of precious metal from domestic sources. From this situation were born numerous state institutions like the military academies and the mining academies, where natural scientists were bound by oath to the lords of the land.

In the France of the Sun King Louis XIV (1643–1715) and his finance minister Colbert (1619–1683), for example, the Paris Academy of Sciences, which was founded in 1666, was expected to cater to the needs of the mercantile society. Under the leadership of this academy the first steps were taken toward subjecting crafts and technology to scientific analysis. In the second half of the 18th century, information of this sort, which has been collected over decades, was published in a multivolume work entitled *Descriptions des arts et métiers* (*Descriptions of handicrafts and occupations*). French scholars of the 18th century were occupied with teaching and research in newly established special schools for military affairs (e.g., École du Génie de Mézières), road and bridge construction (École des ponts et chaussées), and mining (École des mines). Similar institutions were also established in other feudal states to meet mercantile needs. In Saxony the Freiberg mining academy, which was founded in 1765, came under the authority of the leader of the "War and Road Building Commission, Mines, etc." named Johann Wolfgang von Goethe (1749–1832); in Berlin the Academy of Sciences, which traced its origins back to the year 1700 to an initiative of Gottfried Wilhelm Leibniz (1646–1716), adopted Leibniz's motto, "Theoria cum Praxi" (Ref. 78, p. 3).

The relationship between mercantile objectives and science is visible above all in the field of mining. Mining, smelting, and coining were all branches of the economy that predated the new science. But the more the mines were worked, the deeper the shafts were driven into the earth, and the smaller the amount of ore found in the rock, the more effective the exploitation had to be and the more important it was to have better water control and better separation processes. Scientific knowledge was very useful in the latter areas. Specialized books and state-supported training took the place of the person-to-person transfer of secret information about specific processes. "As the ground yielded its treasures less generously than before, it became clear that new methods would be needed both in the supply and the manufacture of products, and only a generation armed with the weapons of science could create new things" (Ref. 158, p. 11).

The practice of science in the first half of the 18th century can be illustrated by the example of the Schemnitz Mining Academy, which was founded in 1735 during the Hapsburg monarchy under Maria Theresa. The purpose of this institution was "to provide the training to candidates on an appropriate theoretical, but at the same time practical basis which will qualify them for officer service in the state mining, metallurgical, and coinage enterprises." The institution was under the "direction" of the office of the county exchequer in Selmecbanya (Schemnitz) and under the general supervision of the imperial exchequer in Vienna, that is, Her Majesty.... Acceptance into the school in any case depended on the approval of the exchequer... The number of teaching departments and professors were fixed beginning in 1770 at three, namely: (a) mathematics, physics and mechanics; (b) mineralogy, chemistry, and metallurgy; and (c) mining." In addition, there was a collection of minerals which contained, according to a description dating from 1830, "splendid specimens of gold and iron ore, lime spar...geological specimens of diagnostic value, particularly of granite, syenite, itacolumite, mica-slate, flinty limestone, marl, the most beautiful formations... [and] crystal models of gypsum." There was also a laboratory, which was praised as being outfitted with really "royal munificence," and a library in which—according to a catalog of 1770—even contained a copy of Musschenbroek's "Essay de Physic." (The first professor of the Department of Mineralogy, Chemistry and Metallurgy, Nicolaus Josef Jacquin, was a Dutchman and had himself attended Musschenbroek's lectures while a student at Leyden.) Limited stipends were available to the students; in turn they were subject to strict rules: the professors were "obligated to keep careful watch over the attendance of the collegians and to report any unauthorized absences to the Finance Office. Such absences are to be punished by the financial officer by withholding the portion of the stipend corresponding to the days in question... ." In the first few years of its existence the total enrollment of the Mining Academy was 3248, including "48 counts, 51 barons, and 325 noblemen... ." Upon completion of their training the graduates of the Academy were eligible for positions in other fields in addition to mining, metallurgy, and the minting of coins. From this pool of eligible graduates were recruited the officials of the mining agencies, the forest managers for the forests reserved for mining, and the officials of the mining finance office and the mining accounting office. Because of their general technical education, however, the graduates were also in demand for appointment to technical positions in other branches of government service... . (Ref. 124, pp. 12ff).

THE ÉCOLE POLYTECHNIQUE

With the continuing emancipation of the middle class and the industrial revolution, in many countries new institutions of science became established as the 19th century approached. In revolutionary France of the 1790s the École

Polytechnique was founded. This was a facility that was much more than just a school for technology. This school was associated with many personalities whose names have been applied to physical or chemical laws or to mathematical principles: Arago, Biot, Cauchy, Dulong, Fourier, Fresnel, Gay-Lussac, Liouville, Navier, Petit, Poisson, and Carnot to name a few. "The premier school of the world, a school which can be called a wellspring with full justification, a school which causes other countries to envy us"—thus Francois Arago characterized the institution (according to Ref. 119, p. 20). And even a hundred years later it was described by the mathematician and scientific administrator Felix Klein as one of the "most important intellectual factors of the 19th century" (Klein, 1927, p. 5).

The direct occasion for the establishment of the École Polytechnique was the disorganization in higher education during the French Revolution, particularly in the field of civil and military engineering. On the initiative of leaders in the École des Ponts et Chaussées and Génie militaire, the École centrale de Travaux publics was established on March 11, 1794, in Paris. This was intended to centralize engineering education and broaden its scope. The next year it was renamed the École Polytechnique. Its organization was patterned after the Schemnitz Mining Academy and the military school of Mézières. Under the monarchy, family and social position governed admission to engineering schools. Now admission depended on entrance tests, which any Frenchman could take who was "well behaved, loyal to republican principles, knowledgeable in the area of arithmetic and the fundamentals of algebra and geometry, and between 16 and 20 years old." After reforms and new starts affecting the higher specialized educational institutions were in place, including the Schools of Artillery Applications, Engineering, Bridges and Highways, Mines, Surveying, and Marine Engineering, it became the mission of the École Polytechnique to provide students with a broad scientific educational foundation for these disciplines. Instruction was in the military fashion: the students were organized into "brigades;" lectures, laboratory instruction, self-study, and recitation and consultation were scheduled just like shop sessions or military exercises. Figure 2 depicts a typical day at the École Polytechnique in about 1810, when it was administered by the Ministry of War and its students looked forward to careers in Napoleon's army. Both students and professors in the new institutions of higher learning enjoyed high social prestige. The publication of the *Journal de l'École Polytechnique*, as well as the requirement that professors publish the text of their lectures, established incentives, a sense of belonging to an elite group, and high standards. Research and teaching, as well as theoretical and experimental investigations, were viewed as belonging together (Ref. 31, pp. 1–43).

FIG. 2. This barracks courtyard scene from the year 1814 shows not a general before his troops but Napoleon before the students of the École Polytechnique. This first technical university of France has not lost its military orientation even to this day. It is administered not by the Education Ministry but by the Defense Ministry. The students wear uniforms and take part in military ceremonies. From this group are recruited most of the leading technocrats in French government projects such as the Atomic Energy Commission (Ref. 115).

Natural science, as practiced at the École Polytechnique, was already quite different from Musschenbroek's nature research a hundred years earlier. The emphasis on mathematics gave rise in many areas of physics to theories which today are part of the foundation of modern theoretical physics. Thus, for example, the mathematical procedures used by Fourier in analyzing the conduction of heat became "Fourier analysis," a useful tool for theoretical physicists, even though today Fourier's concept of heat must be characterized as outdated. Technical problems were the starting point for many investigations. Navier, a student at the École Polytechnique, for example, was sent to England by the French government in 1821 and 1823 to study the methods used there for building iron bridges. Up to that time stone was the preferred building material. Bridge-building engineers of the 18th century supported theories and experiments which could provide information on the maximum load limit of these materials. Navier, who since 1819 had been assigned at the school of bridges and highways to lecture on the strength of materials, after his trips to England stressed that for building metal bridges it was not enough to know the maximum load limit. An additional important value to be determined was the modulus of elasticity. (The modulus of elasticity was first introduced into mechanics by Thomas Young in 1807, but in a manner different from that of Navier. Today, however, Navier's definition is generally used.) He himself made tests on iron, which he later used to build the Invalides Bridge in Paris. Following the new principles for teaching and research, Navier was not satisfied with the practical analysis for bridge-building purposes. Starting with concepts from molecular theory, he developed differential equations for the elasticity of a three-dimensional isotropic (homogeneous) body, which were later extended by Cauchy to anisotropic media (crystals) and developed into a general theory of elasticity and optics. Like Navier, Cauchy was also a student at the École Polytechnique, then at the School of Bridges and Highways. Both, after some years as practical engineers, were appointed professors at the École Polytechnique.[194]

The model of the École Polytechnique was also imitated in other countries. In the United States the West Point Military Academy was organized according to this pattern and became one of the most advanced American institutions for teaching and research in the 19th century. In Europe polytechnic institutions were established mainly in the German states and from these later grew many institutes of technology, such as Karlsruhe (1825), Munich (1827), Dresden (1828), Stuttgart (1829), and Hannover (1831): "Cradles of a science-based technology—especially since the middle of the century—which provided strong support for the industrial movement inspired by the middle-class culture (p. 160).

"TRAINING AND EDUCATION" IN GERMANY

At the same time, starting in Prussia, a number of university reforms and new starts took place, which gave German universities a new moral code for the conduct of scientific work: scientific endeavor should be carried out without regard for any outside objective but for its own sake alone. This was the requirement of the majority of educational reformers closely associated with an idealistic, neohumanist intellectual mind set. Freedom of teaching and of learning were fundamental principles of the new universities. Human education without ulterior purpose was a duty of the state—with this reform program education was placed under the official protection of an Education Ministry set up specifically for this purpose. Career opportunities were created, and an associated system of competition among scientists with respect to research performance, a "research imperative,"[196] was established. "Publish or perish," the current guideline for getting ahead in today's scientific community, could also have been the advice given by the Prussian Ministry of Education in the 1820s to a young instructor with his sights set on a career as a university teacher. In the judgment of some historians of science, this policy caused "a scientific revolution, a revolution not so much in scientific thinking as in science as an occupation, as a sociological phenomenon" (Ref. 121, p. 226). In education-conscious Prussia, with its idealistic set of guiding principles for the practice of science, with its "research imperative," and with decentralization and competition in the university system, the social structure for the further development of science as a professional activity was put in place this way. Since the ideals of this science policy ran counter to "polytechnism, or whatever one might wish to call the actual trends" (Ref. 119, p. 3), as a philologist wrote to the founder of the Karlsruhe Polytechnic in the year 1930, an antagonism rose between universities and institutes of technology that lasted into the 20th century. Nevertheless, the association between "education and training," between "intellect and industry," or however one might express the contrast in slogans, increased the range of professional opportunities associated with scientific and technical activity.

LOBBIES FOR SCIENCE

A criterion for judging the speed with which science in the 19th century became a "sociological phenomenon" can be found in the appearance and growth of scientific societies. In 1822 a number of scholars, most of them court and military physicians, met in Leipzig to establish the Society of German Natural Science Researchers and Physicians (Gesellschaft Deutscher Naturforscher und Ärtzte, GDNÄ). While the academies of the absolutist period remained under the influence of the lords of the land, the appearance of this society marked the start of the kind of science to be seen in the age of the middle class. The GDNÄ was conceived as a counterorganization to the Imperial Leopold-Caroline Academy. One sees in both the person of the principal initiator and in the circumstances surrounding the establishment of the GDNÄ that citizen's science had to be organized almost as an "alternative movement" against the constraints of the prevailing absolutist system.

> Especially in the person of Lorenz Oken we have a politically involved "convinced politician," who stood up for his ideals of freedom and unity in temperamental to fanatical fashion and who was an uncomfortable subordinate for even the most liberal authority... . In 1811 Rostock University declined to appoint Oken because he was "a hot-tempered man who disre-

garded the boundaries of proper behavior." In 1819 he had to quit his job because of differences with his two bosses, Minister Goethe and Archduke Karl August of Saxony-Weimar... . In 1827 Oken was given a professorship in physiology at Munich. Here too, differences with King Ludwig I cropped up and the latter wanted him transferred to Erlangen... . Politically one could place him somewhere in the vicinity of the anarchists. He became the founder of a new type of scientific organization, more because of his political than his scientific activity. The place where the society was founded was also significant. At that time Leipzig was the great German fair city after Frankfurt and a center of literary activity in Germany. It was located in one of the few German states where freedom of the press and freedom of assembly were allowed.

The society Oken founded was at first a society focused on meetings. Membership was basically open to "all persons scientifically engaged with the study of nature and medicine" (paragraph 6 of the Statutes of 1822). To become a member, one must have earned a doctorate degree and in addition must have been active as a writer. The GDNÄ thus was based on a rather broad foundation measured solely on the basis of scientific qualification...."meetings take place yearly. The site of the meetings changes" (paragraph 9, paragraph 10). This new idea of a moving society obviously corresponded to the needs of the scientific community of thosc days for communication and information (Ref. 137, p. 254ff).

The growth of GDNÄ membership and the formation of specialized subject sections (Tables I and II) are evidence of the rapidly increasing organization of natural science taking place at that time and its differentiation into a series of subject-oriented disciplines. Following the model of the GDNÄ, lobbies for science were also formed in England (British Association for the Advancement of Science), France (Congres Scientifique; Association Française pour l'Advancement de la Science), and the United States (American Association for the Advancement of Science). In addition to these societies a number of specialized associations were established, which catered to the specific interests of individual disciplines. In Germany between 1847 and 1913 65 discipline-oriented societies in the field of natural science and medicine were established, of which at least 55 were more or less closely related to the GDNÄ. Thirty-eight of them split away from the mother society during this time period. At the same time new positions were also created in the universities. Specialization in new subject areas offered new opportunities in the battle of scientific competition. The farther the process of differentiation progressed, the smaller became the role of the natural science research organization as compared with the discipline-oriented societies. Representatives of the GDNÄ made a virtue of necessity when they redefined its role at the beginning of the 20th century.

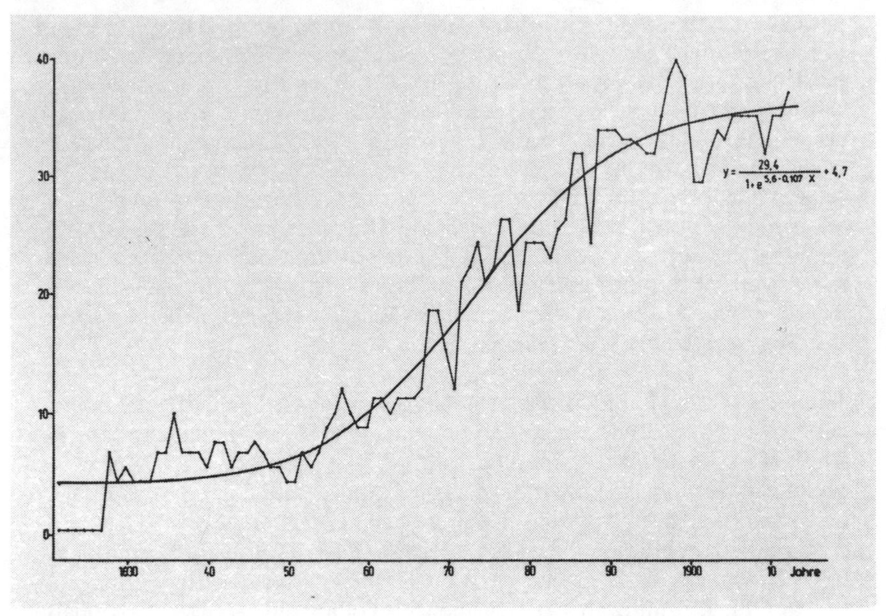

TABLE I. The increasing specialization of natural science in the 19th century can be seen in the growing number of professional sections in the Society of German Natural Scientists and Physicians (GDNÄ).

> Precisely because of this sudden rise in the number of specialized congresses, the Society of German Natural Science Researchers and Physicians gained increased importance. Why? Because it was the only meeting place where the individual disciplines had the opportunity to talk to one another and it was possible to concentrate the experience gained in the special disciplines and at the specialized meetings (Ref. 137, p. 289).

But this only emphasized that the age of the "natural science researcher" was over once and for all. The future belonged to the specialists in the natural sciences.

INDUSTRIALIZATION AND SCIENCE

"There are three people with whom a scientist must work: his patron, his colleagues, and his public" (Ref. 10, p. 39). In the age of absolutism the researchers still depended directly on the contemporary land owners, who anticipated benefit in the pursuit of these economic and military objectives, even when they just left the choice of favorable days for doing business to their court astronomers. A

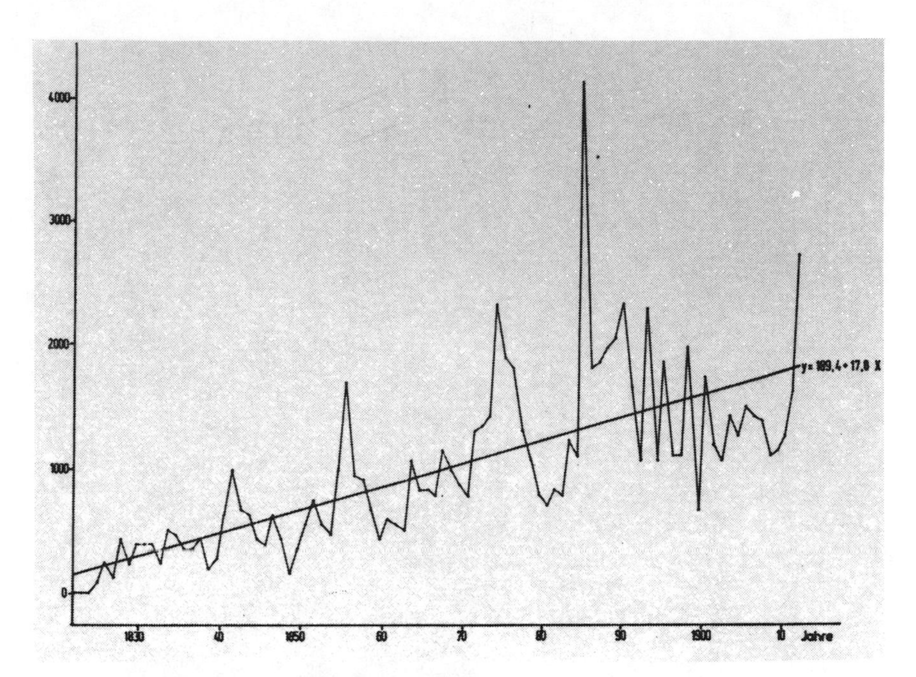

TABLE II. The wide swings in the curve showing the increase in the number of participants shows that the development of the society of natural scientists was subject to changing interests and did not take place without stresses.

famous natural scientist gained increased respect for his patron, and another important function was the amusement of the court. Thus the famous mathematician Leonhard Euler (1707–1783), appointed to the Berlin Academy by Frederick the Great (1712–1786), built dancing water devices for the court, which, however, according to the king, did not function. Electricity, with its unexplainable effects, was eminently suitable for the edification of the nobility (see Fig. 3). How exhilarated must Louis XV (1710–1774) of France and his court have been when Jean-Antoine Nollet, a clergyman who had devoted himself to physics and was the educator of the royal princess, demonstrated a glass vessel invented in Leyden. Filled with water and held in the hand, the "Leyden jar" could store electricity. In a demonstration of this, more than 100 unsuspecting soldiers would join hands to form a discharge circuit for this first condenser, and at the moment when the circuit was closed, it is reported that everyone jumped into the air in unison as the shock hit them. Even in 1801 a professor was complaining that he could never get to his research into electricity because he had to put on demonstrations for the family of the prince.

There was little direct exploitation of the natural sciences for economic

FIG. 3. In the absolutist era the entertainment of nobility was one of the principal responsibilities of the natural scientist at court. Here a boy is suspended in the air by silk cords for purposes of electrical insulation and then is given an electrical charge before the astonished viewers by means of a glass tube. An inquisitive lady tries with her index finger to coax a spark from the tip of the youth's nose.

purposes before the industrial revolution. Processes and procedures used in mining and the manufacture of textiles as well as the theoretical information needed for this purpose were taught in the corresponding specialized schools. Practical experience rather than science dominated preindustrial production methods. In this period, however, the revolution in social relationships pulled down the traditional barriers and assisted in the rise of a production methodology which employed the natural sciences as a weapon in the battle of competition starting in the late 19th century.

A decisive factor was the elimination of the medieval guild system, which sought to forcibly prevent "the conversion of the master craftsman into a capitalist by limiting the number of workers which a single master craftsman might employ to a very low maximum" (Ref. 120, p. 327). "Similarly he could employ apprentices only in the craft in which he himself was a master. The guild vigorously opposed any encroachment of merchant capital—the only free form of capital—which threatened it" (Ref. 120, p. 380).

After the guild restrictions were removed, it was possible to employ many workers in one factory. A second fundamental change gave the workers an incentive to go to work in the factories: the abolition of serfdom liberated a large mass of the population, which, lacking property, was forced to sell its labor. "The direct producer, the worker, could only take charge of his own person when he was no longer chained to the soil and no longer the serf or servant of another person... . Thus appeared the historical trend which changed the producers into workers for wages, freeing them from servitude and the power of the guild, on the one hand, ..., but on the other, these newly liberated workers became sellers of themselves only after having been robbed of all their means of production and all the guarantees of survival offered by the old feudal mechanisms" (Ref. 120, p. 743).

The methods of production used in the factories at first hardly differed from those of the guild shops. But where many performed identical work it soon became clear that more effective use of different skills could be achieved if the products were produced by division of labor. The specialization led to simplified handling procedures, which, because of continuous repetition, could be optimized and carried out with the help of machines. Many machines in a factory, initially operated manually, gradually came to be driven by a central power source. The quality of the products was still to a large extent dependent on the skill of the individual worker. In the competition for market share, however, consistently high-quality and large production figures were demanded. Both requirements were met (first in the production of weapons) by increased automation and the development of more complicated machines: a task for engineers, who since the beginning of the 19th century had been trained in polytechnic schools.

IRON AND STEEL

Science could be called upon to assist when unforeseen difficulties cropped up. Occasions for this arose as iron and steel began increasingly to take the place of the traditional manufacturing raw material, wood. The properties of the new materials were investigated with chemical and physical methods, and in this regard it was possible to reach back to early experience gained in bridge construction. Materials-testing institutes were set up, like the Franklin Institute in Philadelphia (1824), which was engaged both in the training of mechanical engineers and in research. One of the first projects took on the urgent problem of steam boiler explosions (Fig. 4). With a materials-testing machine built specifically for this purpose the strength of the steam boiler materials (iron and copper) was studied at different temperatures.

FIG. 4. The Murz locomotive after a steam boiler explosion in the year 1849. Ever more frequent accidents due to defective materials led to the establishment of materials testing institutes, which were also given research responsibilities.

In Europe also, industry learned to make use of the newly established technical and scientific disciplines to increase productivity and to develop completely new products. Natural laws were no longer applied in a random fashion to production, but were deliberately introduced into the production process or even studied so that they could be introduced. The traditional industries, like the metals industry, were at the forefront of this movement, where industrial processes were subjected to scientific analysis.

Sadi Carnot, a student at the École Polytechnique, in 1824 published his "Reflections on the Motive Power of Fire." In that publication he concerned himself with the study of steam machines, "For they are enormously important and their applications are growing day by day. They appear destined to effect a great revolution in our civilization... . If some day improvements in heat engines progress far enough to make them less costly to build and to fuel, they will make possible an industrial growth of dimensions we can hardly fathom in advance."

In spite of the variety of work reported on heat engines, and in spite of the satisfactory state which they have now reached, very little progress has been made on the theory of these machines, and attempts to improve them move along almost by accident (Ref. 23, p. 4).

Long before heat was recognized as a form of energy, Carnot detected the main outlines of the law which is known today as the Second Law of Thermodynamics. Ignored at first, Carnot's ideas were embraced ten years later by B. P. E. Clapeyron, also an École Polytechnique student and later a professor there. He made them more easily accessible through the use of a new kind of graphic representation and mathematical formulation. Now Carnot's theoretical reflections on the cycles are still having an effect on industrial practice: they became the basis for the development of refrigeration machinery and internal combustion engines.

Further evidence of how science came to be used by industry is seen in its application to the improvement of steel production. In the 1850s, Henry Bessemer (1813–1898), an engineer with a knowledge of metallurgy and chemistry, invented a simple process for converting iron to steel. The important chemical processes involved, however, were largely not understood. Efforts to put the process into practice failed, and Bessemer was suspected of being a charlatan. Then science came to the rescue. In 1859 the chemist Robert Wilhelm Bunsen and the physicist Gustav Kirchhoff at Heidelberg University demonstrated that the breakdown of light by prisms can be used to analyze the chemical elements. In doing this they took advantage of the fact that chemical elements in the gaseous state emit light with a characteristic color. The light emissions accompanying the Bessemer process were a clear invitation to use the new technique of "spectrum analysis." A student of Bunsen's systematically studied the flame in an English iron works and determined the best conditions for producing steel with the Bessemer converter.[195] The consequence was a steep rise in steel production, for example, at Krupp from 5000 tons in the year 1861 to 50,000 tons in the year 1865.[66] All the larger steel producers converted their production to the new process, which had been optimized by spectrum analysis. The demand for chemists rose, and with it the number of students and, in Germany, government support for new faculty positions. Direct participation of industry in research activities, however, remained the exception until the end of the 19th century, since industrialists were not interested in scientific knowledge *per se* but only in its profitable application. Only when effective patent laws guaranteed them exclusive use of their own research results did industrial firms build research laboratories as a competitive weapon.

ELECTRICITY

In contrast to the traditional industries, the newly established chemical and electrical industries were closely tied to scientific research. These new industries did not have to face the problem of replacing traditional ways of doing things with methods based on scientific investigation.

The first application of electrical current, telegraphy, had established itself by the middle of the 19th century as a profitable industry. Copper wires connected the European commercial nations among themselves and with their colonies. The telegraphic connection of the continents of Europe and North America, an enterprise important for world trade, became an international project in which scientists from a wide variety of disciplines studied problems of insulation, mechanical load-carrying capacity, electrical cable loading, signal reflection, and so on. Early experiments failed, and the physicist William Thomson (later Lord Kelvin) was able to assign responsibility for the failure to differing electrical resistance in different sections of cable. The British government then contracted with the chemist August Matthiessen, who operated a private laboratory in London, to study individual cable sections in detail. He determined that the different resistances were caused by impurities, and he developed a method of monitoring the purity of copper, which today has become one of the standard methods in use. When, with the aid of this method, the telegraphic connection (Fig. 5) of the Wall Street (New York) and London City Stock Exchanges was accomplished in 1866, the event was celebrated as one of the greatest achievements of the 19th century.[181]

The expansion of the telegraph industry laid the groundwork for the rapid growth of the electrical industry in the last two decades of the 19th century. On the one hand, its profits provided capital for establishing large production facilities, and on the other hand, it led to the establishment of electrical engineering as a profession and the founding of industry groups such as the English Telegraph Association and the Electrical Engineering Society, organized in Berlin in 1879. The number of those employed in the electrical industry in Germany rose from 1292 persons in the year 1875 to 18,704 in 1895 (Fig. 6) (Ref. 26, p. 53).

Those branches of science concerned with the study of the properties of solids were particularly strongly affected by the electrical industry. The development of filaments for light bulbs spurred the search for materials with the highest possible melting temperature and at the same time being readily workable. Established firms set up research facilities at the turn of the century, such as the "Bell System" (founded in 1875 in the United States), whose laboratory was to become in the 20th century one of the largest centers for studying the properties of solids. The first physicist was hired there in 1880 "to have made

FIG. 5. A marine cable was laid amid great difficulties and dangers (steel etching, 1888). Because of the economic benefit expected from rapid transmission of information between America and Europe, the participating firms put up with failures and high capital losses at the time.

FIG. 6. View of the manufacturing hall for countercomponents in the Siemens-Schuckert Works at Nuremberg. The hall was built in the last decade of the 19th century. The picture was made in the year 1912.

under his direction such special experiments as he may be directed to make, and report on the same" (Ref. 91, p. 519).

THE IMPERIAL INSTITUTE OF PHYSICS AND TECHNOLOGY (PHYSIKALISCH-TECHNISCHE REICHSANSTALT)

The closer together nations were brought by technological innovations like the railroad and telegraphy, the louder became the cry in the new electrical industry for unique, precisely defined electrical units. In the year 1881 the International Commission on the Definition of Electrical Units agreed on a unity of resistance which corresponded fully to a proposal by Werner von Siemens. The work required to define the "ohm" precisely was carried out in England, however, since no laboratory in Germany was able to undertake the costly precision work required. Germany indeed was second only to the United States as an exporting country, but because precision machines had been neglected there, precision measuring devices had to be imported from England and France.

Both the precision mechanics and the optical industries as well as scientific institutions like the Calibration Standards Commission of the Prussian State Survey, the Hydrography Bureau of the German navy, and the Astronomical Observatory complained about the inadequate quality of optical and mechanical instruments and demanded an institute for the "improvement of scientific mechanical devices and instruments." As a result of resistance from established science and from several politicians, the negotiations for setting up an institute of this sort in Prussia dragged on for a year. As the discussions moved forward, it was decided that the institute should serve the entire German empire. Hermann von Helmholtz, Professor of Physics at the University of Berlin, and Werner von Siemens demanded that scientific problems generally should be investigated in addition to technological problems. Siemens, in his letter to the Ministry of Education, stressed the importance of an institute which could concentrate on research without teaching responsibilities. He emphasized the special advantages of this for the nation:

> Important material and intellectual advantages would accrue to the Empire from a scientific laboratory. In the current vigorously fought battle of competition among nations, this country has a decisive superiority in that new paths are trod first and the industries to be based on them are built up first... . It is not scientific education but scientific performance which gains a nation the place of honor among civilized peoples.... (Ref. 74, p. II).

Siemens backed up his request for a research institute by making available to the state a piece of ground in Berlin on which to build it. In 1881 the way was finally cleared for establishing the Physikalisch-Technische Reichsanstalt (PTR). Two divisions were set up, a Technical-Mechanical Division and a Physics-Scientific Division. The PTR, particularly its Scientific Division, became the forerunner of state-supported research facilities, in which the costs of research were shifted to all taxpayers while sparing the profits of the businessmen. It became one of the first and most important centers for the studies of the properties of solids, as an institution in which the interests of science and industry were joined.

While the Technical-Mechanical Division became a testing authority for industrial products on a fee basis, the Scientific Division carried out studies which were mainly commissioned by industry and government organizations. Thus, during the early years of the PTR, physicists developed a method of measuring temperatures of metals up to the melting point, a task assignment originating from the steel industry. The German navy ordered magnetic measurements on iron and steel, since the iron used in ships and torpedoes interfered with the ships' compasses. A reliable light standard was developed for the German Association for Gas and Water Professionals, and the German Association of the Beet Sugar Industry ordered the construction of polarization-measuring instruments to permit determination of the concentration of sugar and alcohol in organic substances.[21]

Foreign reaction provides a basis for estimating the importance of the PTR in the economic and scientific battle waged by Germany for international power and recognition. Similar institutions patterned after the PTR were set up in the mightiest industrial nations: the National Physical Laboratory in England in 1900, the National Bureau of Standards in the United States a year later, the Laboratoire d'Essai in France, and in Japan the "Riken" Physics and Chemistry Institute in 1917.

THE "GÖTTINGEN ASSOCIATION"

Another kind of collaboration of science and industry at the end of the 19th century is seen in the "Göttingen Association for the Advancement of Applied Physics and Mathematics." Felix Klein, Professor of Mathematics at Göttingen University, because of his relationship with the Prussian Ministry of Education (Friedrich Althoff) and with influential industrialists like the Director of the Bayer Dye Works and Delegate to the Lower House of the Prussian Legislature, Henry Theodor Böttinger, became a sort of gray eminence for the organization of science in Germany at the turn of century. After travel to the United States (to the Chicago World's Fair in 1893 and to the 150th anniversary celebration of Princeton University in 1896), he understood how to use the argument of international competition to bring German industrialists to the point where they were ready to provide financial support for applied science. First and foremost, the chemical industry showed itself ready to contribute: the Institute for Physical Chemistry was founded at Göttingen University in 1896 with Böttinger's support. A year later industry contributions made possible the establishment of a Technical Division in the Physics Institute. In 1898 industry support of science at Göttingen University was institutionalized in the form of the "Göttingen Association" (Göttinger Vereinigung"). In the proceedings of the founders' meeting it is stated: "An Association of industrialists meets to provide the resources required by the Physics Institute of Göttingen University to expand teaching... and research in the field of applied physics" (Ref. 119, p. 174). In the same year the Association made money available for an Institute of Applied Electricity.

Research at the new "applied" institutes typically was concerned with the border areas between purely scientific and purely technical disciplines (e.g., electrochemistry, electrometallurgy, stress analysis, hydrodynamics, aerodynamics, and the theory and application of magnetism). The overlap of long-separated disciplines (physics at universities and technology at institutes of technology) was the consequence of science, economics, and government coming closer together (Fig. 7). The pattern of interaction can be described briefly as follows: industry spent money, the government created recognition and new positions for researchers, and science became oriented in the direction of applications. In Germany this pattern became the model for support societies and organizations in the 20th century (e.g., the Kaiser Wilhelm Society), where interdisciplinary research could be carried out on a large scale.

THE FIRST RESEARCH LABORATORIES IN INDUSTRY

By the end of the 19th century it was clear that the natural sciences could make important contributions to technology, but the number of natural scientists employed by industry remained small. Siemens in his memoirs mentioned the ineffective patent law as one reason for this, "...one of the biggest obstacles to free and autonomous development of German industry;" since in the second half of the 19th century patents still lasted for a maximum of 3 years, and since "even for this short period they offered only inadequate protection against imitation, they were as a rule only applied for to obtain concrete evidence of an invention" (Ref. 180, p. 234).

Siemens himself participated actively in the drafting of a patent law, which was passed in a first version in 1877 for the entire country. As in other industrialized countries, it was years before additions to the law guaranteed effective

FIG. 7. Cartoon from the year 1908 marking the tenth anniversary of the Göttingen Association. Industrialists and scientists approach one another. The industrialists hand over a small portion of their money and receive knowledge in return. The almighty Althoff (Prussian Minister of Education) reigns over all. Felix Klein, founder of the Göttingen Association, illuminates the scene as the sun, while the lesser role of Wilhelm II is indicated by the moon.

protection against competition from one's own inventions. Only this guarantee made investment in industrial research worthwhile, as a prominent engineer from General Electric stated:

> Were it not for the patent system the industrial research laboratory would be nonexistent, for no corporation, however wealthy, could afford the great expense of research if its results, as soon as they were commercially devel-

oped, could be copied by competitors at the bare cost of reproduction (Ref. 12, p. 79).

The research laboratory of the Siemens works grew out of a small physics and chemistry laboratory housed in a makeshift building (referred to locally as a "doctor dungeon") in Berlin-Siemensstadt. Its beginnings extend back to the last years of the 19th century, when the chemist W. Bolton (1868–1912) was looking for materials to replace the delicate carbon filaments of the Edison light bulb. Edison's method of searching around the world for better materials was recognized as not the best way to attack the problem. Light bulb manufacturers hired investigators to solve the problem scientifically in well-equipped laboratories. In the Laboratory of Physics and Chemistry at the Siemens plant, Bolton worked his way systematically through the Periodic Table of the elements looking for a material which would show at the same time a high melting point and a vapor pressure that was as low as possible. Tantalum seemed like the most suitable metal, but up to that time it was known only as a powder oxide. After years of work, tantalum filaments were successfully produced, which assured the firm of competitive advantages for some years.

The number of light bulbs sold climbed steadily to almost 10 million by 1911, but in the following years fell drastically to 0.5 million. The reason for this loss of market share was that about 20 scientists and engineers at the laboratory of the General Electric Company had produced a cheaper and longer-lasting filament out of tungsten, which soon came to dominate the market. Thirteen different processes were tested and almost a million dollars were paid for the rights to various filament patents of foreign firms. The head of the General Electric Research Laboratory, Willis Whitney, who himself participated in the development of the carbon filament incandescent lamp, had recognized that only the establishment of a larger research division would provide a basis for successfully competing with developments in Europe. He hired as one of his first scientists William D. Coolidge, who was teaching at the Massachusetts Institute of Technology. He offered to double his salary and assured him that he could devote half his duty time to his own research.[12] The General Electric Company tried very early to create good conditions for research. Thus the laboratory had a scientific library which in 1914 already had 1400 technical and scientific books as well as subscriptions to 64 journals.

To increase the effectiveness of the research conducted at the German Siemens plants and to avoid duplicate work, the in-house patent lawyer, who had very close contact with the laboratories of the firm, requested as early as 1903 that a separate research division be established. At first this idea was opposed by most of the plant superintendents, almost all of whom were members of the board of directors.[39] The progressives were able to get their way only in 1913. These included Bolton's successor, the Göttingen instructor Hans Gerdien (1877–1951), and Wilhelm von Siemens (1855–1919), a son of Werner's. Gerdien was assigned the task of planning a research center. He took advantage of the experience of academic laboratories by visiting in 1914 the famous Dutch university institutes in Leyden and Amsterdam, and the Physics Institutes of the Universities of Freiburg, Heidelberg, Aachen, and Bonn. At the request of Wilhelm von Siemens even the Krupp and I. G. Farben firms opened the gates of their research facilities, although a firm's secrets were ordinarily carefully protected.[179,25] In the summer of 1914 construction began on a generously proportioned research building, which, however, was interrupted by the shift of the Siemens plants to war production. It could be completed only in 1920.

The members of the academically trained staff of the industrial laboratories were usually found as a result of the good relationships that the laboratory leaders maintained with their colleagues in the universities. This was not the least of the reasons why Gerdien continued to teach at Göttingen, and he set up a great staff by "...making contact with the Göttingen institutes... . In this way I could get to know early the new generation trained there and always hire the best staff for the research laboratory" (Ref. 57, p. 1).

This was one way to transfer continuously the latest state of academic research to the industrial laboratories. A second way was via common projects shared by industrial physicists and their colleagues from government research facilities, who were happy to take advantage of the technical facilities available in industry. Often only industry had the facilities available to produce suitable samples for study—particularly of alloys and very pure materials. The industrial scientists knew exactly which professorial contacts would be productive.

For example, a physicist from the Siemens laboratory recommended in 1913: "...as I had the opportunity to report a while ago, I have noticed that the Auer Company and the AEG are supporting Professor Lummer in his work through the building and donation of equipment, etc., evidently in order to keep up to date and possibly to take advantage of new technical ideas. I think it is not impossible that new angles on arc lamp technology are emerging or have already emerged. Would it not be in order for Siemens and Halske also to link up with Professor Lummer in some manner, perhaps by sending over someone from the Laboratory of Physics and Chemistry" (Ref. 140)?

Before World War I the German electrical industry was a leader in the application of scientific methods to production. The economic upswing before the First World War placed the chemical and steel-working industries in a position to establish research laboratories or expand those that already existed. Krupp, for instance, restructured its Research Division in 1909 by setting up its Physical Research Institute in a new building and dividing it into Metallurgical, Metallorgraphic, Mechanical, and Physics Divisions. Several scientific workers were hired, among them a physical chemist who had worked at Göttingen with the most famous investigator in metals research, Gustav Tammann.[51]

At the end of the year 1911, *Scientific American* announced a series of articles entitled "Made in Germany" which was concerned with the introduction of science into German industrial laboratories:

> The announcement began, "One hears it stated repeatedly that Germany is the most scientific country in the world, scientific not only in the fact that it has produced the most important chemists, physicists, biologists, and physicians, but scientific to the further extent that it developed a system for making commercial use of the knowledge stored in the minds of its professors. By publishing the articles announced here we hope to show American businessmen that a chemist, a physicist, or a bacteriologist can mean more for the growth of his business than a new machine, a reduction in the pay scale, or a drop in transportation costs." In the eight articles that followed topics were discussed like "A Laboratory for Manufacturers," "Wealth Through Effective Use of Raw Materials," and "Get Rich with the Help of Scientifically Trained Personnel" (Ref. 173).

In the United States industrial research has been closely associated with the name of Alexander Graham Bell (1847–1922), who founded a substantial sector of the telephone and telegraph industry. Bell's enterprise grew within a decade from the small Bell Patent Association founded in 1877 to the giant American Telephone and Telegraph Company (AT&T). In 1907 the research activities of AT&T and the Western Electric Company (which had been attached to the Bell empire as the production shop) were combined. In 1911 "fundamental research" was defined officially as a component of the organization's program.

> "Fundamental" studies in connection with transmitters, receivers, duplex cables, telegraphs, and particularly repeaters, which were to be investigated "on the broadest possible lines." "The increasing number of problems intimately associated with the development of the telephone business, which require especially exhaustive and complete laboratory investigation... . To make adequate progress," the new branch "should include in its

personnel the best talent available and in its equipment the best facilities possible for the highest-grade research laboratory work" (Ref. 91, p. 533).

In order to find suitable staff, the Bell managers turned to the renowned physicist Robert Millikan (1868–1950), who at this time was occupied with electron discharges in high-vacuum tubes. In his autobiography Millikan recalls the visit of a Bell representative:

> He started our conversation as follows: "Mister John J. Carty, my chief, and the other higher-ups in the Bell System, have decided that by 1914, when the San Francisco Fair is to be held, we must be in position, if possible, to telephone from New York to San Francisco. ...We want you to help us in this job, as follows: let us have one or two, or even three, of the best young men who are taking their doctorates with you and are intimately familiar with your field. Let us take them into our laboratory in New York and assign them the sole task of developing the telephone repeater" (Ref. 91, p. 532).

Millikan sent several of his doctoral candidates to AT&T. Within three years the new division grew to more than 40 employees, including at least seven scientists with doctoral degrees.

The American Association for the Advancement of Science also took an interest in the matter of industrial research. In 1914 it established a Committee on Research in Industrial Laboratories, which included members from both industry and academia. The Committee discussed subjects like "Organization of Industrial Research," "Selection and Training of Students," and "Cooperation between Industry and Universities" (Ref. 5, p. 34). In the same year the Canadian government appointed a council of advisors for scientific and industrial research that included, among others, the Director of the Physics Laboratory of the University of Toronto and the President of the Steel Company of Canada.

In the First World War a new area of endeavor was opened up to natural scientists as industrialized countries began to search for substitutes for raw materials that were no longer available. In addition, they made themselves useful in the development of new and more effective military technology and in raising the level of productivity in the manufacturing of war materials. While in peacetime expenditures for research, considered as costs, were to be held as low as possible, this economic arithmetic yielded to the demands of war. Industry important to the war and industry that could shift its production to war material could more than offset the decline in private demand through government contracts. Expenditures for research were correspondingly modified to fit the new market conditions. Gerdien had the following to say about the laboratory of the Siemens firm: "It soon became clear that, in view of the probable long duration of the state of war, a total shift to tasks related directly or indirectly to the war effort had to be the only conceivable course of action, and from the viewpoint of the Fatherland as well as in the interest of our organization it had to be the only requirement" (Ref. 56, p. 49).

With the newly established and newly organized industrial research laboratories a new career field was opened up to graduates of the institutes of technology. For the first time there was a large-scale alternative to the limited number of positions in the institutes of technology and in the government research facilities, and to the option of a teaching career in higher education. In 1918, for example, the chemistry and physics laboratories of the Siemens plants employed about 100 persons, half of whom had a diploma from an institute of technology. On the other hand, it was incumbent upon the institutes of technology to meet the increasing demand for physicists in industry.

MATERIALS RESEARCH

Industrial research, by virtue of its mission, was concerned with the properties of the materials used by industry. As the electric power network grew, large amounts of energy became available regardless of location. Higher pressures and temperatures could be reached, and the replacement of the steam engine by the internal combustion engine and the electric motor permitted the operating speed of machinery to rise. The resulting increased loading placed on the materials used in machinery and equipment made it more important to know their mechanical properties such as hardness, brittleness, elasticity, and strength. At the beginning of this century, therefore, attention was focused on finding out about these properties. Many years of experience had taught that tiny admixtures of foreign materials would affect these properties. Steel for railroad rails, for example, was hardened with small additions of carbon and magnesium. Without understanding the processes in detail, various additives were systematically tested up to the mid-1920s. The results often depended on chance, as was the case in 1909 when there was an attempt to develop a light-metal alloy for manufacturing cartridge cases that would be cheaper than the brass cases used up to that time. At the Center for Scientific and Technical Studies set up by the weapons and munitions manufacturers of Germany, a temperature-treated aluminum-magnesium alloy was investigated over a period of time. In the course of this study it became clear that the hardness of the metal had increased with the passage of time as a result of aging (Ref. 75, p. 275). This

FIG. 8. The bracing of rigid airships (here that of Zeppelin LZ129) was fabricated from the light metal duralumin.

material, which was called "duralumin," of course, was not suitable for cartridge cases, but the "age-hardening" process became important for light construction, particularly for struts in zeppelins (Fig. 8) and in airplane fuselages. Duralumin replaced the usual wood in propellers and was used in pistons for both aviation and automobile engines.[2]

At Osram's Research Laboratory there was a systematic search, by adding chromium, nickel, and carbide-forming elements, for a hard steel to use in draw dies for making wire for light bulbs. The result was a steel hardener called "Widia" ("dia" as in "diamond"), which was manufactured by Krupp and introduced for the first time in 1927 at the machine-tool exhibition in Leipzig. This made it possible to increase the cutting speed of tools and to work with some materials for the first time.

The microscopic structure was ignored in these investigations, although there was certainly reason to relate the properties of raw materials to their atomic structure. As early as 1914 Theodor von Karman (1881–1963), assistant to Ludwig Prandtl (1875–1953) in Göttingen, summarized the comprehensive observational information from materials-testing institutions in a paper on the physical basis for strength of materials theory. They investigated mainly phenomena that affected the risk of fracture of materials (such as mechanical

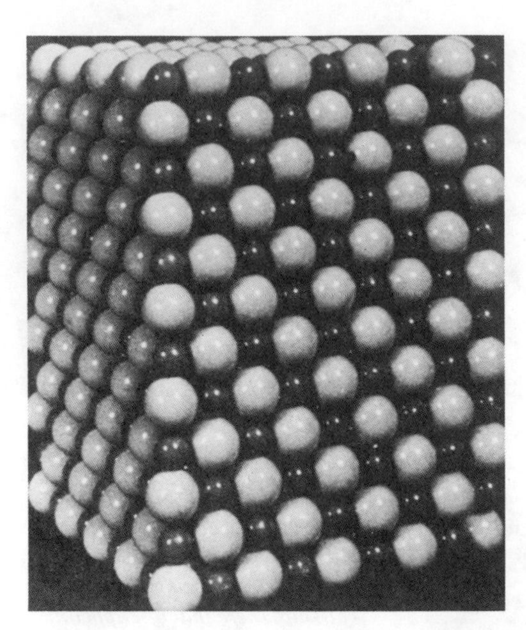

FIG. 9. Model of the crystal lattice of table salt. The sodium and chlorine atoms are arranged in strict regular fashion. In a monocrystal, this orderly arrangement extends over the entire crystal.

hysteresis and aftereffect phenomena). They reached back to a model developed by Prandtl, which was based on a perfect regular crystal, a so-called monocrystal (Fig. 9). Prandtl's calculations had shown that the observed phenomena could not ocur in such a crystal. But the metals actually used in practice are made up of many small crystal grains, the so-called microcrystallites (Fig. 10), which are separated from one another by grain boundaries. When a material is overstressed, it breaks along the grain boundaries or other interruptions in the crystal lattice.[137] At first Pandtl found little reaction to his model. It was too theoretical for application in industry and physicists had only limited interest in the strength of materials. Hence he published the mathematical formulation of his model only in 1928, at a time when "physical strength research through the study of crystal structures and also from another direction through atomic physics has again become fashionable (in the case of pure physicists it is becoming really fashionable, perhaps for the first time)" (Ref. 155, p. 87).

One basis for the change lay also in the use of x rays for analysis of materials, which grew after World War I (see p. 60). This method can be used to study distortions of the atomic lattice by elastic and plastic deformation as well as the process of age-hardening, the structure of mixed crystals, and the phenomenon of metal fatigue. The study of metals produced by different industrial processes showed that different processes caused different orientation of the crystal grains and thus different properties. X-ray diffraction provided information on the processes occurring in the alloying of metals. It showed where in the atomic

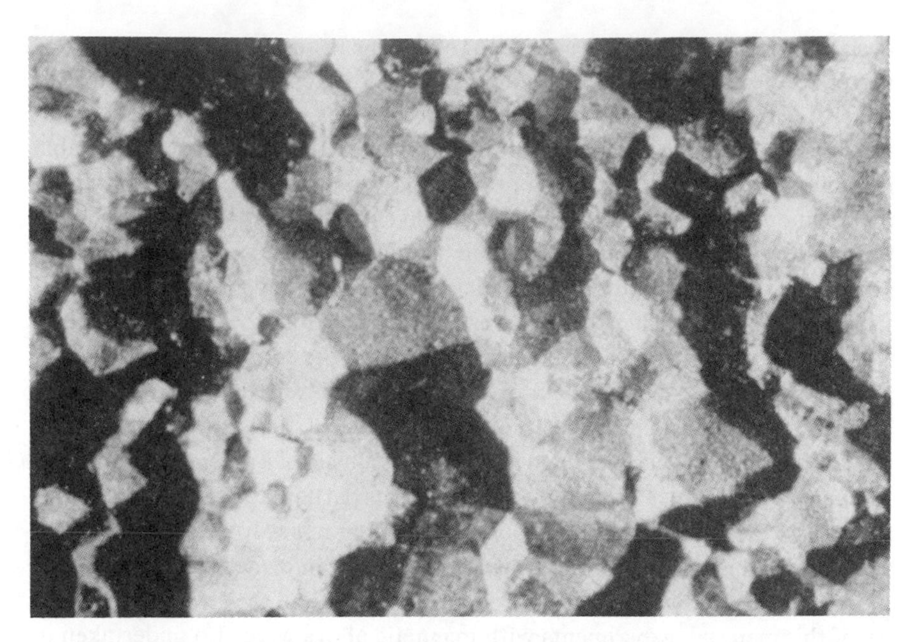

FIG. 10. Fracture surface of a nickel-chromium alloy. One sees clearly how the mass of metal is made up of numerous crystallites joined together.

lattice the added alloy materials were lodged and how the distance between individual atoms was changed by the addition of foreign atoms. In this way it became possible to make alloys with desired properties.

> No one doubts that we are living in an age of alloys, since every day marks the birth of several new ones with particularly useful properties... . It is astounding, however, to realize that until recently the whole process of producing alloys was a hit or miss proposition... . Two molten metals were mixed and cooled. The result was an alloy. But like a genius born of ordinary parents, alloys often possessed unusual properties reminiscent of neither side of the parentage (Ref. 46, p. 78).

One branch of industry especially interested in the "unusual properties" of alloys was that to which the telephone and telegraph companies belonged. The transmission of high-frequency signals over long lines was severely impaired by attenuation. At the beginning of this century Michael Pupin, an immigrant from a province of the Austro-Hungarian monarchy and later Professor of Electrophysics at Columbia University in New York, found a solution and sold his patent to the American Telephone and Telegraph Company. By appropriate adjustment of the resistance, capacitance, and inductance of the cable to the desired signal frequency he was able to reduce the losses through reflection. He increased the inductance of the whole cable by adding coils (Pupin coils) spaced widely apart (at intervals of a few kilometers). This method, of course, could not be used for underwater cables. In the latter case the inductance of the cable itself had to be raised. One option provided for wrapping the conductor with a metal tape in which an external magnetic field could build up a magnetic field on the inside which was as high as possible (high magnetic permeability). In the search for a suitable material, iron–nickel alloys were investigated in the Bell Physics Laboratory. These usually were used for resistance wires. It was found that the physical properties depended not only on the composition but also on the method of production. The magnetic properties were particularly affected by the rate at which the molten material was cooled. The first result of these studies was an iron–nickel alloy, which was the subject of a patent application in 1917 and whose name "Permalloy" pointed to its particularly striking property, its high permeability. Permalloy found application not only in the wrapping of submarine cables but also in transmission circuits, tranformers, filters, and (as a pressed powder) as magnetic cores for coils.[37] Because of the diverse applicability of this material, experiments with magnetic alloys were also undertaken in European and Japanese industrial laboratories. The purpose was to achieve similar properties in an alloy with a different composition, both to get around the American patent and to use as little of the costly raw material nickel as possible. An alloy of iron, silicon, and aluminum with high permeability was produced in Japan in the 1930s. This nickel-free material became especially important during World War II, when supplies of nickel were very limited.

SUPPORT OF RESEARCH AND GREAT POWER POLITICS

Pictures of the world's fairs, where since 1851 the products of the industrialized nations were placed on exhibition at irregular intervals in the metropolises of the West, illustrate clearly the unparalleled growth of the capitalist states. In the words of Napoleon III (1808–1873), the world's fairs were "not simply bazaars but shining manifestations of the strength and genius of peoples." A look inside the Krupp pavillion at the Chicago World's Fair of 1893 shows what such "manifestations" looked like (Figs. 11 and 12). The competition among imperialist states to divide up the world was wide open. The talk was of market outlets, spheres of infuence, and deposits of raw materials. The competition of the industrialized nations extended to all areas which appeared suitable for demonstrating power and prestige—and these included, in addition to economic and mili-

FIG. 11. The Krupp pavillion at the 1893 Chicago World's Fair Exhibition.

tary strength, science as well. The following statements, written in Kaiser Wilhelm's Germany five years before the outbreak of World War I, can be found in similar form in the other great powers of the era as well.

> Military power and science are the two strong pillars of German greatness, and in accordance with its glorious tradition, it is the duty of the Prussian state to see that both are preserved... . The other great civilized nations have recognized the signs of the time, they have in recent years made enormous expenditures in support of scientific research... . This fact is already disastrous from the viewpoint of national policy and will become ever more so in the economic sphere as well. It is disastrous from the viewpoint of national policy because, in contrast to earlier times, with the extraordinarily heightened nationalism that prevails today, every scientific research result bears a national imprint (Ref. 28, p. 47).

In writing of the initiatives of the "other great civilized nations" in support of science the author was referring to the establishment of a series of noteworthy new institutions and endowments, which around the turn of the century were usually financed by private contributions, mostly from industrialists who had become enormously rich in this stormy phase of imperialism. The Carnegie

FIG. 12. At the same exhibition the giant American electrical firm, the General Electric Co., signaled the growing importance of electricity with a giant electric light bulb atop a column.

Institution of Washington, for example, which was founded in 1902, had an endowment of $10 million, which came from the earnings of the American steel magnate Andrew Carnegie and represented only an estimated 1% of his fortune. It was Carnegie's declared purpose to help overcome "our national poverty in science" and to "change our position among the nations" (Ref. 111, p. 69). In conjunction with other industrialists like Rockefeller, Vanderbilt, and Stanford, who personally financed institutions of higher learning, laboratories, the salaries of scholars, and other "charities," Carnegie set an example for newly rich Americans of how they could add national and cultural prestige to their material riches by supporting science. "Endowments for research are becoming an attractive object of private charity," stated the President of Harvard University at the turn of the century (Ref. 111, p. 70).

THE KAISER WILHELM SOCIETY

The example of the large endowments in the competing industrial states—in addition to the American endowments one might mention those of the French scientist-industrialist Pasteur (1822–1895), the Danish brewery owner Carlsberg, and the Swedish explosives industrialist Nobel (1833–1896)—was a challenge to the "greatness of Germany."

> "This is what has happened in other countries; what is happening here at home? It can and may not remain thus; German science, and with it the Fatherland—its internal strength and external prestige—should not be allowed to suffer more serious damage. We need research institutes, not one, but several, established according to a plan and joined together as the Kaiser Wilhelm Institute for Scientific Research." Thus argued the science lobbyist, church historian, and theologian Adolf von Harnack in 1909 in his commemorative paper on the founding of the Kaiser Wilhelm Society (KWS). And again: "Your Majesty further may be kind enough to direct an appeal to the nation and call upon all the wealthy to support willingly Your Majesty in looking after the needs of science, to work effectively together with the State, and to move soon toward the establishment of an association for the advancement of science in the manner outlined above" (Ref. 28, p. 477).

This paper was preceded by efforts over many years to establish outside the universities research facilities where there were no teaching responsibilities. The chemists had tried since 1905 to have established, for their discipline as well, an imperial institute similar to the Imperial Institute of Physics and Technology of the physicists. For this purpose an Imperial Institute of Chemistry Society was founded by prominent university professors such as Emil Fischer, Walther Nernst, and William Ostwald, and representatives of the chemical industry, mainly from the coal-tar dye makers Bayer, Baden Aniline and Soda Factory (BASF) and Agfa, which were later merged into I. G. Farben. By 1910 the Society was able to collect about one million marks in contributions.

These efforts coincided with the long-standing expansion plans of the University of Berlin. Harnack wrote his paper, with the intent of winning the favor of the Kaiser as a prestigious inducement for contributors, at the time of the 100th anniversary celebration of the University of Berlin. Considering the enthusiasm of Wilhelm II (1859–1941) for progress, his agreement could be expected, and on April 7, 1910, Chancellor Bethmann Hollweg (1856–1921) was able to begin compiling a list of potential contributors; he requested the imperial permission to "...identify as quickly as possible wealthy persons...and...initiate limited confidential solicitations in an effort to obtain contributions for the

Kaiser Wilhelm Research Institute and the Imperial Society for the Advancement of Science..." (Ref. 28, p. 49). In October 1910 the Kaiser announced at the 100th anniversary ceremony in Berlin the establishment of the society named after himself. The organizing session and the issuance of the by-laws took place in January 1911. The solicitation of contributions was unusually successful. By 1914 more than 200 private contributors, mainly from the banking and raw materials industries, assembled starting capital of 12.6 million marks. The government made land available as well as staff positions for the Kaiser Wilhelm Institutes, which were established with this capital in rapid succession starting in 1912. Influential positions on the management panels of the Kaiser Wilhelm Society were made available to contributors. Thus Gustav Krupp was honored with the post of Vice President: he joined the Society with a contribution of 1.4 million marks. The membership of the most important panel of the Kaiser Wilhelm Society, the Senate, according to the by-laws, consisted of 15 representatives of business and 4 scientists. The management of each Kaiser Wilhelm institute was entrusted to a director, usually a prominent university professor, who was given wide latitude in the conduct of his scientific work.

In the case of the Kaiser Wilhelm Society we see that, even before World War I, what Harnack had called for in 1890 was achieved on a large scale: "...Like the big power concept and big industry, big science is...a necessary element of our cultural development...Big science requires operating capital like big industry..." (Rev. 28, p. 49). Now science had so much "operating capital" available that criticism was heard not only from the Left:

> "I speak of the Kaiser Wilhelm Foundation. I consider such a pooling of capital, ostensibly for the benefit of science, by the grace of mammon to be out-and-out dangerous. Indeed the power of the purse conquers all, science has been made too much beholden to the Caesar of mammon." To these comments by the Social Democrat delegate Ströbel, Liebknecht added the following: "We are always ready to accept money from the propertied classes—of that you can be sure; we certainly don't resist. But the matter in question here is somewhat different. My friend Ströbel has said it is dangerous to let institutions of this sort be financed by private capital under circumstances in which the goals of the institutes in question and the nature of the activity within the individual institute are influenced by the financing, or at least could be so influenced." Harnack himself, a few months after the founding of the Kaiser Wilhelm Society, asked that thought be given to the fact that "scientific activity must certainly and inexorably become the vassal of capitalism and the associated brutal politics of self-interest, if the state does not keep the situation under control" (Ref. 28, p. 61).

Harnack's statement should not cause one to lose sight of the fact that in Germany the government's share of research support was traditionally higher than in other industrialized countries. At the turn of the century new institutes were established at many universities. These benefited mainly the natural sciences and with them started the development of the universities into large-scale operations (Ref. 20, p. 66ff).

INFLUENCE OF THE FIRST WORLD WAR

Considering that science was viewed as a means of demonstrating national greatness, it is understandable that research support prior to World War I often benefited branches of science where prestige-enhancing results could be expected. Carnegie wanted above all to support the "exceptional man," so that his money flowed into well-tested and established branches of science like geology and geophysics, where American science already had a good reputation and the "exceptional man" was likely to appear earlier than in new research fields like atomic physics. The first two Kaiser Wilhelm Institutes were dedicated in 1912 to the field of chemistry, for which Germany was famous. This was surely a reflection not only of the prestigious character of the discipline but also of the heavy involvement of the chemical industry during the startup phase of the Kaiser Wilhelm Society. Although support based on prestige considerations did not exclude the possibility that sciences with prospects for application would be funded, it still seems that in most cases the opportunity for direct exploitation of research results was not the basic motivation for the contributors.

The situation changed with the advent of the First World War. Because of military requirements, research oriented directly toward applications came to the fore. The Kaiser Wilhelm Institute for Physical Chemistry, for example, was used for military purposes (mainly for the production of poison gas) to such an extent that its staff was increased to several thousands and the director of the institute, Fritz Haber, requested the establishment of a new Kaiser Wilhelm Institute for Applied Physical Chemistry and Biochemistry, to take over the military assignments and free him again to devote more time to fundamental research.[19]

In England after the outbreak of the war, one had to face the fact that some branches of industry, particularly those which depended on scientific processes, had fallen behind the Germans: synthetic dyes and medicines, photographic developers, chemical products, optical glass products, and other industrial goods which had been imported form Germany before the war were suddenly scarce. The shock of the outbreak of war was followed by the shock of realization of German superiority in science, which was viewed above all as "the result of organized professional effort." A secretary of state from the Education Ministry responsible for research matters described the situation thus in a parliamentary debate in 1915: "If this was has taught us one lesson, which we ought to lay to heart more than any other, it is that we have more to fear from scientific organization (...) than anything else" (Ref. 197, p. 194). In the summer of 1915 the British parliament acted accordingly and commissioned an advisory panel consisting of prominent scientists and representatives of industrial sectors heavily engaged in research to develop recommendations for the "(i) instituting specific researches; (ii) establishing or developing institutions or departments for the scientific study of problems affecting particular industries; (iii) the establishment and award of research studentships and fellowships" (Ref. 197, p. 210). What was meant by "specific researches" and "particular industries" became clear by the next financial year (1915–1916): Twenty projects were supported, mainly research on the optical properties of various materials, the properties and composition of alloys, the corrosion of nonferrous metals, the solidification and melting of crystalline substances, and so on. These were preponderantly topics directly related to industrial exploitation. Most projects were assigned to professional institutions like the National Physical Laboratory, the Faraday Society, the Iron and Steel Institute, and the Institution of Mining and Metallurgy. Depending on need, the advisory panel delegated the responsibility to specialized committees, which again were composed of members from the relevant professional societies. Between 1916 and 1926 the Department of Scientific and Industrial Research (DSIR), where this kind of research support had become institutionalized, grew substantially. It took over full responsibility for the National Physical Laboratory, and established additional special research institutions to meet program requirements (the Fuel Research Station, the Radio Research Station, the Low-Temperature Station, and the Chemical Research Laboratory). The First World War was also the backdrop for the establishment (1916) in the United States of a government organization, the National Research Council (NRC), to promote research for "national security and welfare." Composed of leading scientists and engineers from universities, industry, and govenment organizations, including the military, it was the declared objective of the National Research Council to foster cooperation among the various research institutions of the country. In Japan also, efforts were made during World War I to organize application-oriented research. In 1916 and 1917 two of the most significant Japanese institutions for solid-state studies were founded, the Institute for the Study of Iron and Steel, which was commissioned by the Japanese army to study development of magnetic steels, and the Institute for Physical and Chemical Research (RIKEN).

The institutions which were founded in many countries as a result of the First World War to support sciences closely associated with practical applications continued on after the war under conditions that varied from country to country, and they usually were subject to further development. The United States emerged from the war as a new, aspiring economic great power, recognizable at first by its large expenditures on research, and toward the end of the 1920s by the scientific results as well. Under the influence of the National Research Council the support of research developed in the direction of direct grants of assistance, mainly to universities. In 1919 the Rockefeller Foundation transferred $500,000 to the NRC to fund postdoctoral fellowships over a period of five years for research in the fields of physics and chemistry. In the same year the Carnegie Corporation provided $5 million to the National Academy, one-third of which was to be spent on construction of a building for the Academy and the NRC, and the rest was to be expended on NRC projects. Before the war only a small part of the Carnegie money went to universities, but now it contributed to acceptance by American universities of research as an objective to be taken seriously in addition to teaching. The Rockefeller International Education Board, the Guggenheim Foundation, and the NRC were the most important sources of money for the heavy transfer of science which took place in the 1920s between the United States and the European physics centers.[111]

SCIENCE AS A SUBSTITUTE FOR POWER

Conditions were not so favorable for science in all the industrialized countries as they were in the United States. In France, for example, support for research was not significantly higher after the war than before. Worth noting in this regard is the situation in Germany, where one might assume at first glance that after losing the war the climate might not be very good for science. The actual situation, however, was not bad at all: expenditures for research in the field of physics were twice as high in the postwar years as they were in 1914! New support organizations were established, and although the endowment capital of the Kaiser Wilhelm Society had been seriously depleted by inflation, several new Kaiser Wilhelm Institutes were started. To understand this situation one must keep two things in mind: inflation, which mainly hurt members of the middle class and their savings, was a boon for investing industrialists, who could help themselves to the rapidly devaluing savings deposits as borrowers, and at the same time get tax advantages for investing in science. Secondly, the government had to take up the role of financier and guarantor if it was to assume responsibility for seeing that—as Haber put it—Germany after losing the war was at least to retain "its seat on the board of directors of the world.[41,43,44]

The interests of industry and government in science can be seen in detail if we look at the examples of the Helmholtz Society for the Advancement of Research in Physics and Technology (founded in 1920), and the German Science Aid Society. The chemical industry again played a decisive role in the Helmholtz Society as it did previously in the startup phase of the Kaiser Wilhelm Society. Carl Duisberg in particular, the principal force behind the merger of the largest German tar color plants into the I. G. Farben firm,[40] committed himself to this project. At first he merely pushed for the establishment of societies to support chemical research: the Liebig Fellowship Association (1916), the German Society for the Advancement of Chemistry Teaching (1918), the Emil Fischer Society for the Advancement of Chemical Research, and the Adolf Baeyer Society for the Support of the Chemical Literature. "...now that chemistry is taken care of, physics cannot lag behind, especially since, as the mother of mechanics and technology, it is involved in all sectors of industry..." asserted Duisberg on November 24, 1920, to Willy Wien, the Munich physics professor, who became Deputy Chairman of the Helmholtz Society as representative of the universities.[29] Duisberg used as his organizational model his chemical societies established earlier. The intent was to recruit a definite group of industrialists through membership campaigns to support a limited area of research. Duisberg did not assume the chairmanship himself, since he did not feel he was "closely associated with the numerous groups who benefit from physics and mechanical technology;" the chairmanship went to Albert Vögler of the United Steel Works. On January 7, 1921, Duisberg wrote to Wien:

> Under the circumstances Vögler is really the only correct choice for the post...Member of the Reichstag, Member of the Imperial Economic Council, and General Director of one of the largest steel works and colliery firms... . [I] am optimistic that Vögler will keep his word and pull together 40 million marks for the Helmholtz Society... . To the extent I could, I have prepared the way for the recruiting effort among the chemists and have already obtained a pledge of 5 million from the last meeting of the I. G. Farben board, but I have not yet accepted, since, as you saw above, I want more, namely, at least 10 million marks.[29]

Other founders of the Helmholtz Society included the representatives of heavy industry, Ernst von Borsig, Emil Kirdorf, and Hugo Stinnes, whose commitment to research was not as great as in the chemical industry; Duisberg complained that

> ...the representatives of heavy industry, who for the most part do not yet have a correct perception of the importance and significance of science for

technology, and who therefore have contributed practically nothing to us chemists for the three societies which we founded." [29]

Obviously heavy industry had come to a "correct perception of the importance and significance of science for technology" at the start of the 1920s, because it contributed the largest share of all sectors of the economy to the Helmholtz Society. This "perception" came at a time when the iron and steel industry, taking advantage of the cheap production costs during a period of inflation, began to renovate their facilities and steadily to increase production at give-away prices. This brought them valuable foreign currency. The same representatives of heavy industry who accepted science in 1920 were also bothered about politics: Stinnes, a confidant of Vögler, who directed a daughter company of the Stinnes firm, and Kirdorf, the "elder statesman of heavy industry," in October 1920, for example, pressured the national government concerning the domestic need of German industry for a cutback in the rate of delivery of "reparation coal" to France. Stinnes' commitment to an aggressive national policy was just as predictable as his efforts to heat up inflation, which was so beneficial to heavy industry.[67]

Also in 1920, the German Science Aid Society (Notgemeinschaft der deutschen Wissenschaft) was founded, which according to the original plans was also to be financed from the economy by an endowment association. Chairman of the endowment association was Carl Friedrich von Siemens. To avoid a competitive situation with the Helmholtz Society, the endowment association was to run a general campaign with two-thirds of the receipts going to the Helmholtz Society and one-third to the endowment association. Shortly thereafter the split was changed to 50–50%, and in 1924 the endowment association became a full component organization of the Aid Society. In contrast to the Helmholtz Society, the Aid Society was intended to support more than just scientific and technical research, and hence industry rapidly lost interest in it. The expenditures of the endowment association for the Aid Society averaged 2–3% of the government contribution—evidence of the reluctance of the business community to support research without the direct prospect of exploitation. For the government the Aid Society was a means of demonstrating the open concerns of the new parliamentary democracy, which were cultural in nature and friendly to science. In doing this the government avoided the risk not only of turning its academic community into proletarians as a result of inflation (which hit hard precisely this layer of the previously well-to-do middle class) but at the same time assured that the scholars would endorse the democratic system. In view of the fact that the collective mind of the "German mandarin" [162] was still loyal to the Kaiser and was having difficulty reaching an accommodation with

FIG. 13. "Professor Thuisko Groller in Berlin, as a sign of his continuing protest, every morning gives the world a couple of resounding slaps." With this cartoon, the satiric magazine Simplizissimus in 1919 described the unhappiness of the reactionary German professors after the First World War.

the defeat in the World War and the new relationships that followed (Fig. 13), this undoubtedly was not an issue of secondary importance for the government. Both societies divided up the research money through specialized committees consisting of prominent professors from institutes of technology, alloting it to individual researchers (usually in specialized institutes attached to the institutes of technology) in the form of assistance in kind or stipends for projects to run for definite periods of time. The self-management principle for science introduced in this way may also be an indication of the struggle of the democratic government for the loyalty of its scientific elite.

The expenditures of the Helmholtz Society and the Aid Society in the area of physics reflect the differing interests of the two societies. The Helmholtz Society, which was closely linked to industry, had practically no interest in supporting research in very fundamental subject areas such as quantum mechanics and nuclear physics. The money dispensed by this group went by preference to areas related to practical applications, like the physics of materials properties. On the other hand, the government, working through the Aid Society, did not restrict itself to the financing of the basic research spurned by the private contributors, but, as a representative of the overall interests of science nationally, also supported projects like x-ray structural studies for the analysis of metals and electron microscopy.

2

MICROCOSMIC PHYSICS—A SCIENTIFIC REVOLUTION

WHEN THE BUSINESS WORLD recognized that scientific research results could be exploited commercially, not only were radical changes made in the way scientific work was done, but research topics were routinely selected on the basis of their possible contribution to the solution of practical problems. In the age of steam engines and giant structures of steel, engineers in materials testing laboratories and polytechnical schools, together with their academic colleagues in the scientific disciplines, concentrated on studies related to heat phenomena (thermodynamics) and mechanics. As the applications of electricity expanded continuously toward the end of the 19th century, questions concerning the nature of electrical current and the structure of matter became high-priority problems for physics. In the Electricity and Magnetism Division of the Imperial Institute of Physics and Technology (PTR), and in experimentally oriented university institutes, the electrical conductivity of metals and alloys was studied much more intensely, of course, in the 1890s than during the 1820s in the scholarly studio of an Alessandro Volta (1745–1826) or a Georg Simon Ohm (1789–1854), when electricity was certainly known as a natural phenomenon but was of no technological importance at all. The microcosm, the world of the tiniest component particles of matter, had long been an object of great fascination for many kinds of scholars. Speculations producing such things as the atomic model shown in Fig. 14 were based, not on information obtained experimentally, but on philosophical considerations. In the new scientific era, however, ideas concerning the microcosm were tied more closely to observational data developed in several disciplines, including chemistry and mineralogy. But only after the economic and political importance of this area had been demonstrated by the invention of telegraphy, light bulbs, various chemical processes, and so on, did fundamental research receive the support it deserved.

In the first three decades of the 20th century a radical upheaval took place in microphysics, and this became the classic example of a "scientific revolution."[116] Without delving very deeply into the lively philosophical and theoreti-

FIG. 14. According to Lucretius, hard material like stone or iron is made up of "ultimate particles" interlocked (left), while he considered "things which are able to produce pleasant sensations" (Ref. 14, p. 33), such as wine and honey, to be made of smooth round structures (right).

cal scientific discussions that surrounded these events, we shall try to convey in what follows an impression of what led to this upheaval and what the consequences were for solid-state physics.

NEW INSIGHTS INTO THE STRUCTURE OF MATTER

Toward the end of the 19th century physicists believed they were near to completing their picture of the physical world. This belief was based on progress with the Maxwellian theory of electrodynamics, which described the phenomena of electricity and magnetism in elegant form. To solve his later famous equations, Maxwell (1831–1879) came up with electromagnetic waves propagated at speeds close to the measured speed of light. He concluded that light itself was an electromagnetic wave obeying the same laws. Hence optics became a part of electrodynamics. Electrodynamics and Newtonian mechanics thus formed the foundation pillars of classical physics, with which it appeared possible to explain all natural phenomena. For this reason the Munich physicist Philipp von

Jolly advised the young Max Planck (1858–1947) not to take up the study of theoretical physics: Nothing new of significance could be expected. There was no doubt on anyone's part that observations which were still unexplainable and various individual contradictions would soon be accounted for in the theoretical structure. The practical applicability of physics, especially within the growing electrical industry, testified to the fact that a practical understanding of the processes taking place in matter had been achieved. But study of industrially important topics like the intensity and spectral distribution of light (e.g., in arc lamps, incandescent lamps, and spectrum analysis for the Bessemer process) and the electrical conductivity of solids at very low temperatures led to physical discoveries which would clearly show the inadequacy of classical physics and would pave the way for the new concepts of quantum physics.

FIG. 15. First use of arc lamps for nighttime construction work, 1854 in Paris. This light source was too bright to use in private dwellings; it was rather more suited to the illumination of public squares, factory halls, and construction sites. A natural limitation to the length of the workday was thus overcome.

X RAYS, ELECTRONS, AND RADIOACTIVITY

A series of observations that were incompatible with classical physics emerged from the study of electrical discharges, which were used for illumination in the middle of the 19th century before use of the incandescent lamp became widespread. The spark that jumped between two closely spaced electrodes following the application of high voltages was suitable for use as a light source where high intensities were appropriate. Arc lamps made public places shine and transformed night in factory halls into working daylight (Fig. 15). Arc, spark, and point discharges could be studied especially well in evacuated glass vessels or glass vessels filled with rarified gases. After the invention of the mercury air pump the pressure in glass vessels could be reduced further. In discharge tubes (Fig. 16) operated at low pressures several investigators observed the phenomenon of fluorescence, which had already been observed in crystals (e.g., fluorite) and liquids (such as quinine solutions). In this case the glow occurred at the glass wall which was closest to the negative electrode, the cathode. When Johann Wilhelm Hittorf, Professor of Physics and Chemistry at Münster, installed a solid barrier between the fluorescing glass wall and the

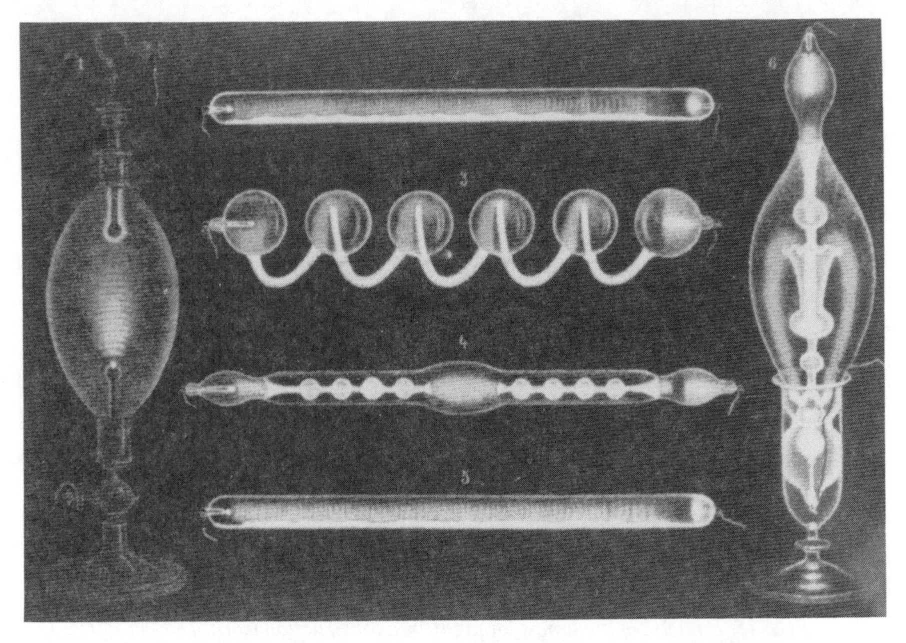

FIG. 16. Various gas discharge tubes from the year 1877. As far as the shapes are concerned, one should remember that the interest in gas discharge was mainly for purposes of amusement.

cathode, a silhouette appeared on the wall behind it (Fig. 17). The phenomenon therefore must have been caused by straight-line, invisible rays from the cathode. Since the physical nature of these rays was not clear, they were called cathode rays because of their origin.

It was not until shortly after the turn of the century, when careful experiments were conducted on the deflection of cathode rays in electrical and magnetic fields, that measurements permitted conclusions to be drawn about the nature of the rays. It was concluded that they consisted of fast-moving, electrically negative-charged particles, which evidently originated from the atoms of the cathode material and possessed a mass approximately 2000 times less than that of the lightest atom, hydrogen. Measurements showed that these particles actually existed in the interior of the atom and could be affected even there by electrical and magnetic fields. The latter measurements were based on analysis of the light emitted by atoms, the spectral lines, as affected by electrical and magnetic fields acting on the atoms. The attempt to describe this particle, called an electron, using theories about the most diverse electrical, magnetic, and optical phenomena, was not successful. The classical laws, which functioned so well in the world of experience up to that time, failed completely in the microcosmic region.

Cathode ray tubes were found in many physics laboratories toward the end

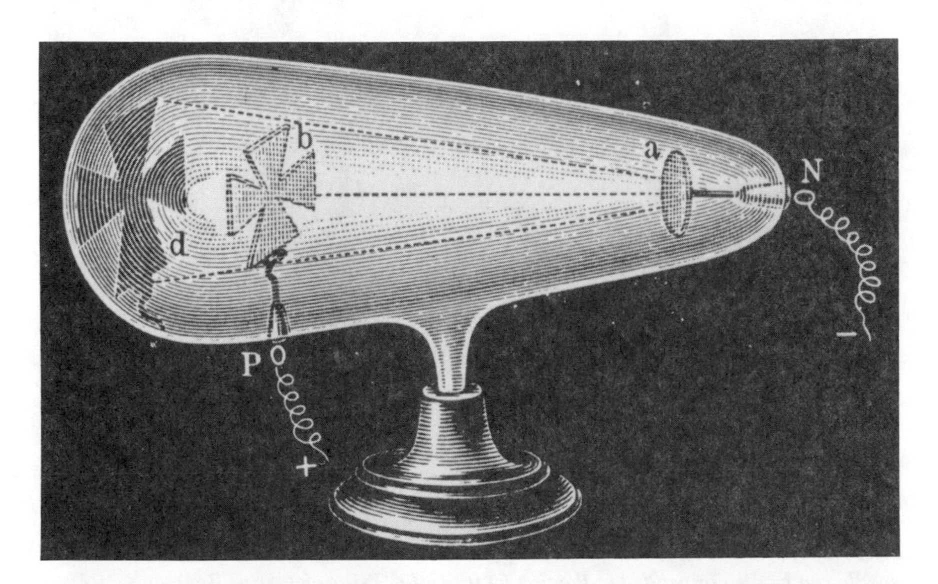

FIG. 17. The silhouette of the Maltese cross on the glass wall of the discharge tube indicated that straight-line radiation was emitted from the cathode.

of the 19th century. Improvements in vacuum technology, as used by the incandescent lamp industry to improve the life and quality of its product, also made it possible to evacuate these tubes more fully. Wilhelm Conrad Röntgen (1845–1923), at that time a still-unknown Professor of physics at the University of Würzburg, experimented with high-vacuum cathode ray tubes of this sort in the 1890s. During his experiments he was surprised in November 1895 to find that a fluorescent screen in the vicinity of the tube began to glow, even though the tube was completely covered with black cardboard. Something was emanating from the tube that possessed such great penetrating power that it could even expose photographic plates through black cardboard. Since Röntgen did not know what this "something" was, he called it x rays. He placed various objects in the path of the rays, and these were penetrated to differing degrees by the new rays. The pictures obtained were truly amazing: they showed the coin in a closed purse, the bones of the hand, and bullets in the barrel of a gun (Fig. 18). Because the effects of these rays could easily be seen even by lay persons, news of them spread quickly to the public and Röntgen's pictures appeared in newspapers

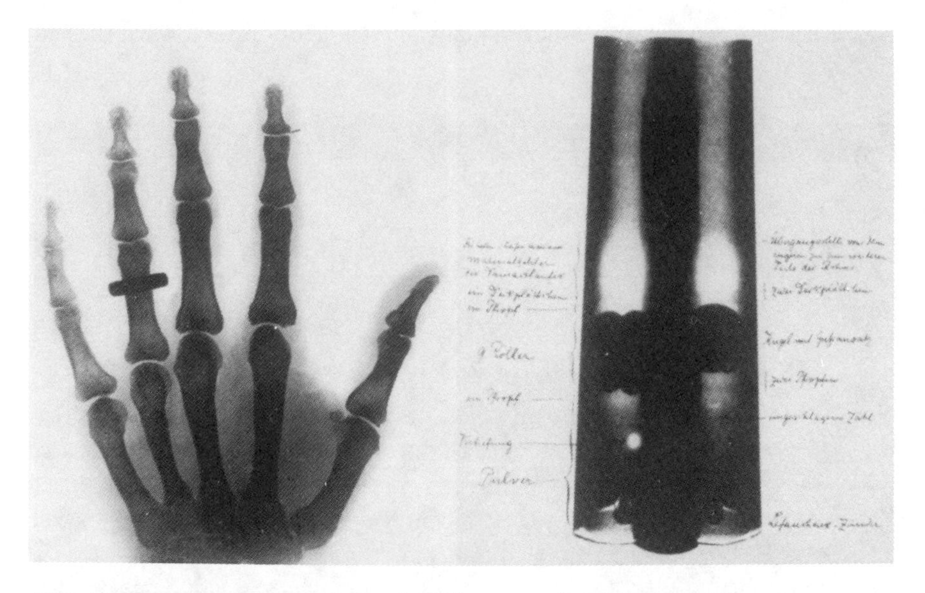

FIG. 18. Left: X-ray picture of a hand which was made a few months after the discovery of x rays. Flesh, bones, and ring were penetrated differently. The end of a needle can be seen imbedded in the first joint. Right: X-ray picture of Rontgen's hunting weapon with his marginal notes from the summer of 1896. We see here already how x rays can be used for materials testing.

throughout the world. Within a few months after news of the discovery became known, many investigators were experimenting with x rays.

At the École Polytechnique Antoine Henri Becquerel (1852–1908) wanted to demonstrate a relationship between the Röntgen rays and the light phenomena in the discharge tubes. He experimented with fluorescent materials from his father's rich mineral collection and used photographic plates to detect the radiation.

In 1896 he discovered that uranium minerals also altered the light-sensitive layer without previous treatment with radiation. He determined that the radiation emanating from the minerals could penetrate matter. The radiation would not respond to external influences and was linked only to the presence of the element uranium. Hence the uranium itself was emitting this radiation. But what was the source of the energy contained in the radiation? It must have come from the interior of the atoms. In subsequent years it became clear that the material emitting the radiation discovered by Becquerel underwent change entirely on its own, without external intervention. Up to that time the immutability of the chemical elements was considered to be a law; only the alchemists had dreamed of transforming one element into another. The search began for more "radioactive" materials. Marie Curie (1867–1934) demonstrated that the element thorium also emitted "Becquerel rays." Together with Pierre Curie in 1898 she isolated by chemical separation processes two additional previously unknown elements, polonium and radium, which also were radioactive. Working with cathode rays, x rays, and radioactive decay, completely unexpected physical phenomena were discovered within a few years. Jolly's prophecy "that nothing new of significance could be expected" did not turn out to be true. Further surprises were waiting, and Max Planck, who did not follow the advice of his teacher, contributed significantly to their explanation.

RADIANT HEAT

The newly discovered invisible rays had already shown that classical physics could not adequately describe the processes occurring in the atomic region. But the first steps toward a new microphysics started, not from the x, cathode, and Becquerel rays, but from routine measurements of the radiation of a heated object at the Imperial Institute of Physics and Technology (PTR). At the instigation of the German Association of Gas and Water Professionals the radiation from glowing metals, particularly platinum, was being measured there for the purpose of developing a reliable light standard. The energy content of the radiation emitted at different frequencies, and thus the energy distribution spectrum, was determined. In his inaugural address Friedrich Kohlrausch, at that time President of the Imperial Institute of Physics and Technology, emphasized the economic importance of such measurements:

> The tangible benefits of such investigations would include, for example, the consistent rating of electrical and gas lamps, hundreds of types of which are submitted for testing every year, with lighting hours numbering many thousands. The economic significance of such work is apparent when we consider that Germany pays lighting costs that are roughly estimated at hundreds of millions of marks every year (Ref. 164, p. 14).

The results of the precise measurements by the PTR deviated from the expected radiation distribution, and especially at long wavelengths the measured intensity systematically exceeded the calculated values. The deviations provided Max Planck with a reason to look into the matter. There was, however, a further basis for Planck's interest in radiant heat. In the case of radiation emitted by an object that absorbs all the rays that strike it (a "blackbody"), the relative proportion of different wavelengths in the radiation does not depend on the material used but only on the temperature of the radiating object. Planck saw in this a useful approach in the search for "absolute" natural laws not influenced by disturbing peculiarities connected with the material involved. The thermal radiation of a blackbody was produced by investigators at the PTR by using a hollow chamber filled only with radiation and heated to the desired temperature. The radiation escaped from the cavity through a tiny opening and could be studied. Planck, who routinely kept in close touch with the physicists of the PTR, was always aware of the latest status of the measurement. In his autobiography Planck described the course he adopted wherein he assumed that a large number of vibrating systems (oscillators) were the carriers of the energy. These oscillators could exchange energy among themselves by sending and receiving electromagnetic radiation. He viewed the oscillators as abstract constructs, as models for calculation, not identifying them with actual oscillating particles:

> Use of the Maxwellian electromagnetic theory of light appeared to be a direct way to a solution. That is, I pictured a cavity filled with simple linear oscillators or weakly damped resonators with different natural periods, and I expected that the exchange of energy among the oscillators through mutual irradiation would in the course of time lead to the steady-state condition of normal energy distribution corresponding to Kirchhof 's law (Ref. 143, p. 23).

After working on this for years, in 1900 Planck succeeded in developing a quantitative description of thermal radiation with a formula which reproduced the experimentally measured distribution of radiation intensity without resorting to any sort of material quantities, thus removing the divergence between measurements and theory. Planck, to be sure, had to introduce a new natural constant which had never before been detected, which he labelled h. Planck's model could have failed with purely classical calculations, because in the classical scenario the oscillators would have released their energy completely to radiation without developing an equilibrium between matter and radiation. Planck could get around this only if he considered the energy of the oscillators to consist of a definite number of finite parts. The magnitude of the energy quanta was h times the vibration frequency of the oscillators. "...that the energy is forced from the beginning to remain grouped in definite quanta was [at that time] a purely formal assumption, and I really didn't think much about it."[142] As far as Planck was concerned, therefore, the introduction of the new constant, called Planck's constant in his honor, in no way marked the establishment of a quantum theory. Einstein (1879–1955) was the first to make it clear that the "purely formal assumption" was the key to a fundamental natural law, when five years later he made a true "quantum constant" out of h. Using Planck's formula Einstein showed that short-wave heat rays behave in many respects like small particles which cannot be broken down further. He postulated the existence of light quanta, packets of radiant energy, as it were, with an energy equivalent to "h times the frequency." Thus the energy of the light quanta depended on the color of the light. As an experimental touchstone for this idea Einstein brought in the photoelectric effect: that is, the release of electrons from a metal by light (e.g., from a sheet of aluminum) should only be possible if the light quanta exceed a threshhold energy corresponding to the work function. In other words, if the magnitude of the light quanta depends only on the frequency of the light, it is not the strength (intensity) of the light but its color which is critical to the release of electrons. It was not until 1916 that precise measurements by the American physicist Robert Millikan provided definitive proof that Einstein's theory of the photoelectric effect was correct. At the same time these measurements provided a new method for determining Planck's constant (h), which was completely independent of heat radiation theory.

The most direct proof for the light quantum hypothesis was provided by experiments conducted in 1923 by Arthur Holly Compton (1892–1962) at Washington University in St. Louis. Compton had worked at the Westinghouse Laboratories as an industrial physicist during the First World War. In 1919 he devoted himself again to basic research. A fellowship from the National Research Council made it possible for him to spend some time doing research in England with the respected physicist Ernest Rutherford (1871–1937) at the Cavendish Laboratory, where there was a long tradition of studying the radiation of radioactivity and its interaction with matter. Back in the United States, Compton investigated the scattering of x rays by metals.

The wave nature of x rays had just been incontrovertibly proved in the previous decade on the basis of diffraction by crystals. It was precisely on this wave property that new procedures for the structural analysis of solids were based, procedures that at this time were coming into widespread use. Compton's measurements, however, did not fit a pure wave representation. The x rays reflected by metals differed in frequency from the incident ray, depending on the angle of reflection. All attempts to explain this finding within the framework of classical diffraction theory failed. With the light quantum hypothesis, however, the "Compton effect" became quite clear. An x-ray quantum striking an electron in the metal transfers more or less energy and momentum to this electron depending on the scattering angle, as in the collision of billiard balls. Hence the x-ray quantum emerges from the diffraction process as a "weaker" quantum, that is, with a lower frequency. This made available an additional new method for determining the quantum constant h. More important, however, was the fact that there now were different experiments in which x rays appeared as particles in one case and as waves in another. This wave-particle duality became an open challenge to the physics community.

SPECIFIC HEAT

When it became possible to create temperatures close to absolute zero (see p. 54), deviations from the classical laws of heat theory were discovered with increasing frequency. Max Planck, looking back in a presidential address in 1913, described this problem very vividly:

> To warm a piece of copper from −250 to −249 °C or one degree did not require approximately the same amount of heat as that required to warm the copper from 0 °C to 1 °C, but some 30 times less. If one were to drop the initial temperature of the copper even more, he would find the corresponding quantity of heat to be even less by several times with no definable limit. This fact is diametrically opposed not only to all normal ways of thinking but also to the requirements of the classical theory (Ref. 141, p. 67).

In 1906 Einstein pointed out a way toward a solution to the problem with the help of quanta. Thus for a second time he advanced the quantum theory, this time with an explanation of the "specific heat of solids," where the question was not about radiation quanta but about the quantification of mechanical processes.

> While Planck's theory of radiation gets to the heart of the matter, we must expect to find in other areas of heat theory as well, contradictions between present molecular kinetics theory and experience, which will crop up along the chosen path and be disposed of (Ref. 33, p. 180).

Einstein suggested acceptance of the oscillators postulated by Planck as carriers of heat in solids. They can take up energy only by multiplying the energy quanta. That is to say, every oscillator can be excited by 1, 2, 3, or more energy quanta to stronger vibration and thus to higher temperature. It cannot, however, take up or release energy, for example, from 2.5 quanta or less than one quantum. Einstein showed that, hence, at a given temperature many vibrations for the specific heat can be inactive, as it were. As the object moves toward lower temperatures, more forms of vibration become inactive, and this explains the observed decline in the specific heat. As a sample calculation for this theory Einstein chose the measured values for carbon, whose anomalously low specific heat had been observed as early as the middle of the 19th century; a deviation from the classical theory was seen even at room temperature. For a long time it was viewed as an exception which was just accepted as unexplained until experiments at low temperatures showed that all the substances investigated behaved in similar fashion. To demonstrate this it was necessary to eliminate the disturbing influence of thermal motion which masked the quantum effects at room temperature. Using low temperatures they could be "frozen out" as it were.

LOW TEMPERATURES: ACADEMIC RESEARCH COMES OF AGE

Many physicists, whose activity was rooted in classical physics, at first closed their minds to the new quantum ideas. Measurements at low temperatures, however, which very often conflicted with classical physics, helped quantum physics to gain increased recognition among the physics community. Physical experiments in temperature regions that were inaccessible by normal methods demanded the greatest care and expensive equipment. Thus the creation of lower temperatures proceeded in step with the establishment of large research facilities. The beginnings of low-temperature physics developed from very practical concerns. Even in the pre-industrial age, as commerce increased and cities grew up rapidly, the cooling of foodstuffs on long journeys and in warehouses became a pressing problem (Fig. 19). In addition, during the 19th century chemists became interested in liquefying gases by lowering their tem-

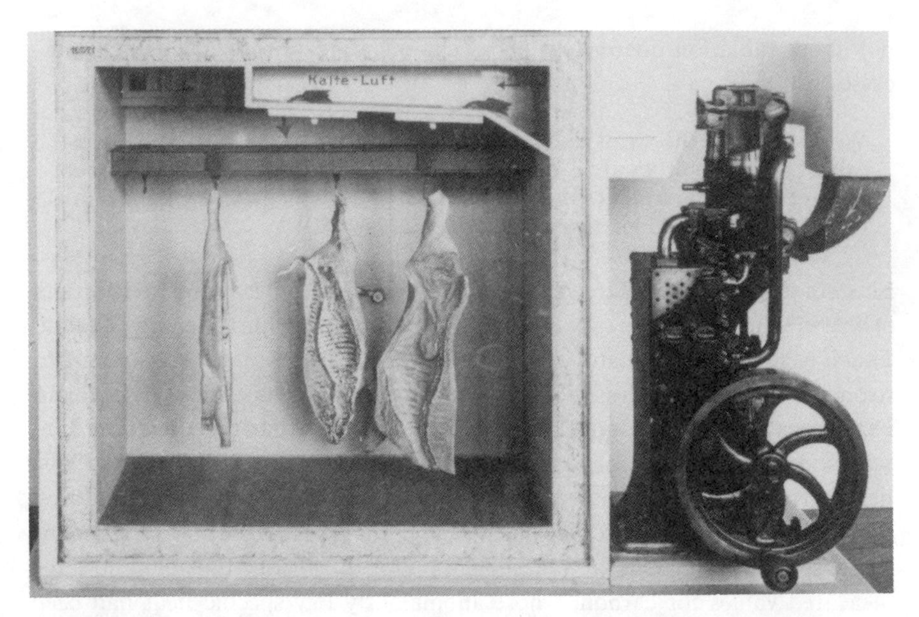

FIG. 19. Low-temperature physics received strong impetus from cooling technology developed for the growing food industry. The picture shows a cold-air machine from the year 1886.

perature, as a simple method for recovering nitrogen and oxygen from the air. These two gases were important in the chemical industry and in steel production.

In 1893 two French researchers observed for the first time the formation of small fog droplets following the sudden expansion of oxygen. Attempts to convert oxygen completely into the liquid state developed into a race toward absolute zero, in which three groups of researchers participated. A year after the fog droplet experiments, K. S. Olszewski (1846–1915) and Z. F. Wroblewski (1845–1888), at the Chemistry Laboratory of Cracow University, collected a few cubic centimeters of liquid oxygen in a device that they had obtained from one of the French scientists mentioned above. The two researchers could not maintain the momentum of this forward leap in the pursuit to lower temperatures, because the poor financial condition of the Galician University forced them to always seek "cheap" solutions to their problems and to spend much time on trivialities. Olszewski, for example, fashioned by hand a bellows out of an old boot and had to make every screw and washer himself.[106]

Circumstances were much better for James Dewar (1842–1932), professor of "experimental philosophy" at Cambridge and professor of chemistry at the

Royal Institution of London, where he worked mainly in the low-temperature field. The laboratory had already acquired a tradition and hence was well equipped. The major portion of the budget,which was contributed by private persons and industrial firms, was available to Dewar for experiments at low temperatures. Many measuring devices were made available on loan by industrial firms for ongoing experiments. With all the compressors, pumps, and liquefaction equipment, the place resembled a machine room more than a chemical laboratory (Fig. 20). Dewar was assisted in the conduct of his experiments not only by his students but also by two permanently employed engineers. It was in this laboratory that Dewar in 1895 succeeded in liquefying hydrogen. In doing this he reached temperatures that were only 15 °C away from absolute zero. In addition to specific heat, Dewar studied other properties of solids at these temperatures, such as the electrical resistance of many metals, thermoelectricity, and mechanical stress. Helium was the only remaining gas which

FIG. 20. The compressors and pumps make Dewar's laboratory at the Royal Institution look like an engine room.

had not been liquefied. The next goal of the low-temperature laboratory was to liquefy this material.

A third cold laboratory was established in Leiden, where Heike Kamerlingh Onnes (1853–1926) occupied the first teaching chair of experimental physics in Holland. The motto of this manufacturer's son was "knowledge through measurement." "It is measurement finally that in the deviations provides the natural material for new hypotheses regarding the properties of the molecule," and hence he wanted to measure as accurately as possible. "A modern physical laboratory therefore must be modeled upon astronomical lines" (Ref. 109, p. 2). Hence he invested planning and organizational effort in equipping his laboratory like none before him. He realized that his research facility with its many modern devices would require a staff of skilled and specially trained assistants. Hence he established a school for instrument makers and glassblowers, which was financed by the Education Agency of the Netherlands (Fig. 21). He was able to enlist financial support for his experiments not only from the Dutch refrigeration industry but also from its foreign counterparts. How thoroughly Kamerlingh Onnes planned his research work can also be seen in the establishment of a scientific journal, the *Communications from the Physical Laboratory at the University of Leiden*, which was devoted exclusively to the

FIG. 21. View of the mechanical shop of the School for Instrument Builders and Glass Blowers of the University of Leiden in 1900. The best graduates worked in Kamerlingh Onnes' Low Temperature Laboratory.

work of the Leiden laboratory. The first issue appeared in 1885. In addition to scientific investigations this journal contained precise descriptions of the apparatus used in the laboratory, so that for several decades it became the bible of low-temperature physics. After several additions to the structure, the Low-Temperature Institute became strongly reminiscent of a factory, and because of its copper tanks and pipes it was jokingly referred to as the brewery (Fig. 22). Helium was liquefied for the first time in 1908. With this achievement, researchers had come within 1.5 °C of absolute zero.

FIG. 22. View of the Low Temperature Laboratory of the University of Leiden, where helium was liquefied in 1908 and the superconductivity of mercury was discovered in 1911. Seated in the middle is Heike Kamerlingh Onnes, and standing behind him is G. Flim, master mechanic and head of the School for Instrument Builders.

SUPERCONDUCTIVITY

Kamerlingh Onnes' research group chose to make measurements of electrical resistance at low temperatures with the simplest experiment they could think of after they had liquefied helium. A theory formulated in 1902 by Lord Kelvin (1824–1907) was championed in the Leiden laboratory. According to this theory electrons should be torn from atoms by thermal motion as the temperature rises. Resistance is created by collisions of electrons with one another. Since electron movement decreases as the temperature is lowered, and thus the collisions become less frequent, the resistance should decline until the electrons finally freeze tightly to the cores of the atoms and no longer contribute to conductivity. Then the resistance should increase sharply and at absolute zero it must be infinite. A minimum should appear at a few degrees Kelvin. After the first measurements, however, at temperatures of about 1.5 K (− 271.5 °C), this idea had to be abandoned—a minimum was out of the question.

The resistance curve rather seemed to resemble that for the temperature dependence of the specific heat. Therefore, Kamerlingh Onnes accepted Einstein's explanation of the specific heat and transferred it to the electrical conductivity. "In particular an obvious assumption to make is that the free path of the electrons, which provide conduction, is determined by the elongation of the Einstein oscillators" (Ref. 105, p. 22). He substituted, in the classical formula for electrical conductivity, the energy of the quantum oscillators instead of the kinetic energy of the electrons. Thus for the first time the quantum concept was brought into the calculation of electrical conductivity, albeit improperly. When he inserted numerical values for the oscillator energies, Kamerlingh Onnes found that the resistance should become zero even at a point above absolute zero. Mercury turned out to be the only metal that still showed a measurable resistance at liquid-helium temperatures (around 4 K or − 269 °C). In spite of serious experimental difficulties in working with this metal, which is liquid at room temperature, Kamerlingh Onnes in 1911 decided to investigate the resistance of mercury.

In the course of their experiments the Leiden research group discovered that the electrical resistance did not drop off gradually to an unmeasurable value (analogous to the behavior of the specific heat), but rather disappeared suddenly at a temperature of 4.2 K (see Table III). The difference in electrical conductivity above and below this temperature of sudden transition was as great as that between an insulator and a metal. Kamerlingh Onnes reported on his measurements at the Solvay Congress in November of the same year and had to admit that more than just the classical theory failed in explaining this unexpected conduction characteristic: his attempt to explain the sudden disappearance

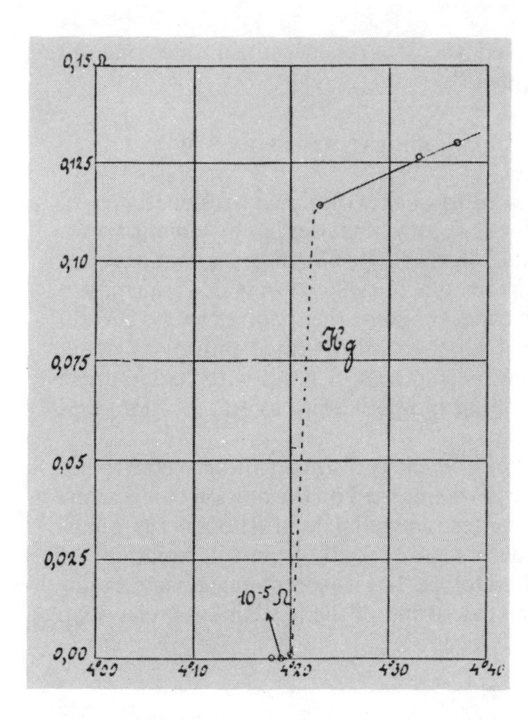

TABLE III. Record of measurements of the electrical resistance of mercury at very low temperatures. At 4.2 K ($\approx -269\,°C$) resistance disappeared abruptly. The researchers suspected at first that a short circuit had occurred. Later it became apparent that they had discovered an unexpected property of matter, which was labelled superconductivity.

of electrical resistance using quanta was also unsuccesssful. It would be about a half-century later before the phenomenon of superconductivity could be successfully described with the help of concepts from quantum theory.

THE FIRST SOLVAY CONGRESS

The meeting addressed by Kamerlingh Onnes was convened in November 1911 with the general theme "The Theory of Radiation and Quanta." It was organized at the instigation of Walter Nernst (1864–1941), professor of physical chemistry at the Univesity of Berlin. In 1906 Nernst had drawn up a general theorem on the behavior of substances at low temperatures, from which it followed, among other things, that the specific heat of solids had to disappear at absolute zero. After a meeting with Einstein it became clear that the "Nernst heat theorem" represented a necessary consequence of the quantum theory. From that time on, Nernst belonged to the circle of dedicated champions of the quantum theory. He succeeded in convincing the industrialist Ernest Solvay, who was strongly interested in natural science, to finance an international meeting of physicists to discuss the "revolutionary restructuring" of the foundations of physics. Since Solvay was a Belgian, he invited the scientists to Brussels (Fig.

23). As Nernst foresaw and intended, the congress became a milestone in the development of the quantum problem.

> The invitations, which were sent to 18 leading representatives of their disciplines, included the following text:
> It appears that we are now in the midst of a revolutionary restructuring of the foundations on which the kinetic theory of matter has rested until now. On the one hand, this theory in its fully developed form, as no one so far has contested, will lead to a radiation formula, the validity of which contradicts all past experience; on the other hand, the same theory gives rise to certain theorems concerning specific heat which are also refuted entirely by many measurements (constancy of the specific heat of gases with temperature changes, validity of the rule of Dulong and Petit down to very low temperatures).
> As has been shown particularly by Planck and Einstein, these contradictions disappear when certain limits are imposed on the motion of electrons and atoms in oscillations around a resting state (doctrine of energy quanta); but this conception again differs so markedly from the equations of motion of material points used heretofore that its acceptance must certainly be accompanied by a broad restructuring of the fundamental views we have held up to this time (Ref. 77, p. 155).

FIG. 23. Participants in the First Solvay Conference in 1911, which was convened at the suggestion of Walter Nernst (first row, far left). The subject of the conference was "The Theory of Radiation and Quanta." The conference was financed by the industrialist Ernest Solvay (first row, third from left). The picture also shows the following physicists mentioned in the text: Madame Curie, Einstein (standing, first on the right), Kamerlingh Onnes (standing, second from the right), Planck (standing, second from the left), and Sommerfeld (standing, fourth from the left).

After this conference, if not before, the quantum concept was no longer just a notion introduced by outsiders. The international discussion of this topic initiated by the elite among the physicists made it a research object of immediate interest. The later architect of the first model of the atom based on the quantum theory, the Danish physicist Niels Bohr (1885–1962; see Fig. 24), was also motivated by conversations with participants in the meeting to study the report of the congress and come to grips with the quantum concept.

THE LAUE EXPERIMENT

One of the speakers at the Solvay Congress was a professor of theoretical physics teaching at Munich, Arnold Sommerfeld (1868–1951; see Fig. 24). In his paper, which was followed by extensive discussion, Sommerfeld reported on the theory of x radiation. His teaching chair at Munich for many years had custody of the "State Mathematics and Physics Collection" of the Bavarian Academy, which provided the theoretician with not only a supplementary emolument as "conservator," but also a mechanic assistant and opportunities for practical experiments. In accordance with this tradition Sommerfeld's theoretical institute was assigned basement rooms in which experimental facilities were housed.

It was there in 1912 that Max von Laue (1879–1960) discovered the diffraction of x rays by crystals: x rays showed evidence of interference after passing through crystals! A photographic plate placed behind a crystal to record the x rays passed by the crystal structure did not show simply a diffuse, weakened spot but rather a pattern of bright points that were located outside the direct line of the incident x-ray beam. Evidently these x rays thus were waves that behaved like light waves and water waves, in that the meeting of crests and troughs

FIG. 24. Sommerfeld (with the cigar) and Bohr established in Munich and Copenhagen after the First World War active centers for theoretical research on the atom. Here ideas about "electron orbits" around the atomic nucleus were developed mathematically and used to explain the spectra.

caused reciprocal reinforcement or extinction of the diffracted bundles of rays. In addition, the experiment demonstrated that the crystal possessed a regular lattice structure which acted as a diffraction grating does for light. This meant the physicist had found a "probe" with which to examine the interior of crystals. Conclusions could be drawn about the lattice structure of crystals from the diffraction pattern. On the other hand, this experiment was such an incontrovertible proof of the wave nature of x rays that Compton's equally incontrovertible discovery of the particle nature 10 years later (see p. 49–50ff) raised the matter of wave-particle duality to a question of the first order. Sommerfeld characterized this discovery as "the most important event in the history of the Institute... . The consequences of this discovery for the understanding of x radiation, atomic structure, and crystal structure are incalculable" (Ref. 30, p. 62).

The impetus for this was provided by discussions about crystal optics problems (such as double refraction) which should be based on the crystal structure. Peter Paul Ewald, who was writing a doctoral dissertation on this subject under Sommerfeld, later recalled a conversation with Laue, at that time an unsalaried lecturer (Privatdozent) in the deparmtent.

> I began telling Laue of my problem... . He asked: "What are the particles, which are regularly arranged, are they the atoms?" I confessed that little was known about this and that they might also be quite large groups of atoms. To his next question: "But what are the distances between the particles?" I had to answer that this depended on the assumptions concerning the nature of the particles, and that assuming their nature the distances could be calculated from the density of the crystal... these distances would be extremely small compared to the wavelengths of the light for which I meant to calculate the law of propagation... .
> Instead of appreciating the expressions of the light field in the crystal, which I wrote down for him, he asked repeatedly: What would these expressions be if the wavelength of light were much shorter than you assume?" (Ref. 30, p. 67).

In speaking of shorter light waves, Laue was talking about x rays, the electromagnetic nature of which had just been discussed in Sommerfeld's department. The relationship between wavelength and lattice distance in the crystal appeared such that Laue expected interference phenomena when the rays passed through the crystal. The incident x-ray beam, according to Laue, should stimulate the crystal atoms to emit their own radiation (fluorescence radiation), and the spatially symmetrically arranged emitters should overlap in an interference pattern. When Laue found that Sommerfeld rejected his idea (and rightly so, because fluorescence radiation is not capable of interference), he secretly ordered that the experiments be conducted. After initial failures Laue, Friedrich, and Knipping actually obtained interference pictures in 1912. The pictures, of course, were not created by fluorescence radiation, but rather the incident x rays were diffracted by the regularly arranged atoms of the crystal, and depending on the direction of the diffraction rays, extinction or reinforcement was observed. The observation of interference points was at the same time evidence for the wave nature of the x radiation and for the regular arrangement of atoms in the crystal.

The discovery, for which Laue received the Nobel Prize in 1914, soon became a widely used method for determining the structure of crystals. It gained importance as well both for microscopic representation and for technical applications in research and the testing of materials.[36]

THE "QUANTUM MAGIC IN THE SPECTRA"

Another category of physical phenomena that gave rise to a revolution in microphysics can be grouped under the heading "spectral lines." The splitting of light with a prism was used as early as the 19th century in industry as a useful method for determining the composition of gases. This method made use of the fact that atoms made to glow by heat or an electric current emit light of a very definite frequency, which is characteristic of each individual type of atom. These spectral lines became the "atomic music of the spheres," when Bohr in 1913 demonstrated a relation to the quantum theory in connection with a new model of the atom (Fig. 25). According to this model electrons revolve in the atom like planets in orbits around the atomic nucleus in such a way that their energy cannot assume any value in continuous fashion but only very definite "quantized" values. A transition from one orbit to another should only be possible if the energy difference is emitted or absorbed as light, or more precisely, as light quanta. Sommerfeld, who took this idea and built it into a comprehensive theory (Figs. 26 and 27) wrote the following in the preface to his book published in 1919 entitled *Atomic Structure and Spectral Lines*:

> The spectra are speaking to us these days in what is truly the atomic music of the spheres... . All integral laws of the spectral lines and atomistics flow ultimately into the quantum theory. It is the mysterious instrument on which Nature plays the music of the spectrum and whose rhythm she uses to regulate the structure of atoms and nuclei (Ref. 9, p. 178).

This flowery description of the subject should not hide the fact that one was dealing here—again using Sommerfeld's kind of language—with "problems which are best solved, not with the head, but with the opposite part of the body" (Ref. 9, p. 176). The relationship of the spectral lines to particular, discrete

FIG. 25. Sommerfeld had this model of the atom made for the Deutsches Museum, Munich. It shows one electron moving in a circular orbit around a central atomic nucleus, corresponding to the Bohr concept of the hydrogen atom.

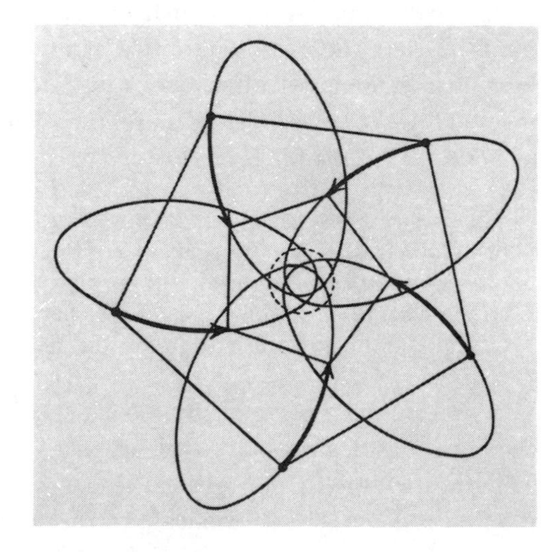

FIG. 26. Sommerfeld extended the Bohr atomic model by postulating, in addition to circular orbits, ellipses as well for the movement of electrons around the nucleus (above). It was possible to explain theoretically many spectral lines from the mathematical description corresponding to this concept.

FIG. 27. Sommerfeld's students later "praised" this accomplishment of their teacher with the "proof" given below.

Proof, that the circle is a degeneration of the ellipse

atomic energy levels and the assignment of "quantum numbers," considering the abundance of empirical data (92 different kinds of atoms; "splitting" of spectral lines under the influence of electrical and magnetic fields), became a laborious struggle with sets of numbers and selection rules which were to be used to decipher the secret of "the quantum magic in the spectra" (Table IV). "This mixture of incomprehensible numerical mystique and undeniable empirical success was naturally for us young students a source of great fascination," wrote Werner Heisenberg (1901–1976) in his autobiography (Ref. 71, p. 55). He had spent his first year of study with Sommerfeld at the beginning of the 1920s. In a letter from Sommerfeld to Einstein dated January 11, 1922, we read

> In the meanwhile I have recognized wonderful numerical rules for line combinations...and have presented them in the third edition of my book [*Atomic Structure and Spectral Lines*]. One of my students (Heisenberg in his third semester!) has even interpreted these rules using a model.... Everything fits, but basically is not yet understood. I can only advance the technology of the quanta, you must build their own philosophy.... (Ref. 76, p. 96).

QUANTUM MECHANICS

When physics is called the "science of the century"[79] and when people refer to the "golden age" (Sommerfeld) or the "heroic time" (Oppenheimer), the reference is usually to the flowering of theoretical physics during the 1920s. Even before that physics produced the "theory of relativity," which caused a sensation. The publicity accorded this theory was certainly in direct contrast to the layman's ability to understand it. The publicity was probably based rather on

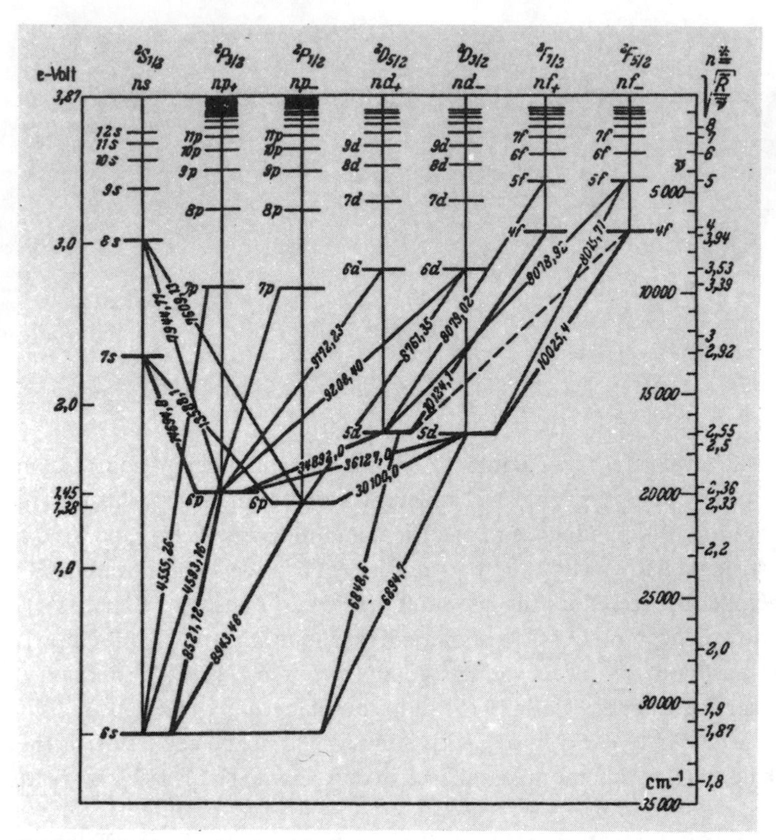

TABLE IV. Energy levels with designation of states in the cesium atom. The transitions between states, which are linked by slanted lines, correspond to the spectral lines of cesium. The characterization of energy states revealed what a complicated science spectroscopy had become.

the idea of "relative" as a term that did not put off ordinary people and the journalistically effective picture of its creator, Albert Einstein. Another theory was born in the 1920s which was not, to the same extent as the theory of relativity, the product of a single genius outsider and hence was less attractive to the news media, but which in its impact on society would be much more significant. This theory was quantum mechanics.

In contrast to the broader concept "quantum theory," which includes the totality of the theoretical interpretations and models at the atomic level since about 1900, the term "quantum mechanics" designates a theory which gave a solid footing to the decades of struggle to understand quanta by the formulation of laws of motion for the microscopic world. Newton's and Einstein's mechanics describe movement under the influence of gravity, like the circulation of the planets around the sun or the oscillation of the pendulum of a clock. Quantum mechanics, on the other hand, describes movement under the influence of atomic forces and permits, for example, the calculation of the energy states of bound electrons in atom and molecules. This calculation is just as necessary to understand the quantized nature of electromagnetic radiation as it is, for example, to understand chemical binding.

"There has rarely been in the history of science a decade as exciting as that from 1923 to 1932" (Ref. 183, p. 3). With these words one of the group of some 50 to 100 physicists who contributed substantively to the development of quantum mechanics begins his scientific autobiography. The new theory grew at sites located mainly in Europe, and a large proportion of them in Germany, where the institutionalization of theoretical physics in the context of university development was now far advanced.

At the beginning of the 1930s it appeared that all fundamental questions on the structure of matter were soluble with the help of quantum mechanics—chemistry, metallurgy, crystallography, and even biophysics and biochemistry were thus given a new foundation. Many electrical properties of metals, semiconductors, and insulators could be explained by it, and given this understanding new technological applications were suddenly shifted to the realm of the possible.

MATRIX ALGEBRA

The breakthrough to an understanding of the various forms in which the "quantum magic" could appear took place in 1925 mainly on two fronts. The solution was called "matrix" or "quantum mechanics" by some, and "wave mechanics" by others. Both sides had their stars: the matrix mechanics advocates included, for example, Max Born (1882–1970), Werner Heisenberg (1901–1976), Niels Bohr (1885–1962), and Wolfgang Pauli (1900–1958), and their names were synonymous with the outstanding groups or "schools" of theoreticians in Munich (Sommerfeld), Göttingen (Born), and Copenhagen (Bohr). As far as the authors of wave mechanics are concerned, those included above all Louis de Broglie (born 1892) and Erwin Schrödinger (1887–1961); both worked alone and were not members of "schools." The paths to solution in both camps were entirely different, but the results obtained were identical. It was very quickly demonstrated that both approaches were equivalent, and since then the term "quantum mechanics" has been generally adopted.

The riddles of the spectral lines provided the critical impetus for matrix mechanics. In 1924 Wolfgang Pauli, who came from the Sommerfeld school as did Heisenberg, discovered a general principle that could be used to explain a series of spectroscopic questions: "In my view it all boils down to considerations of statistical weights, in the sense that in certain appropriately arranged stationary conditions no more than *one* electron can exist" (Ref. 81, p. 183). In this way Pauli informed his teacher Sommerfeld of his "exclusion principle" on December 6, 1924. He jokingly named this principle the "housing office for electrons." A physics student today learns the "Pauli principle" as follows: "Only those states of the atom actually occur in nature in which any two electrons differ in at least one of the four quantum numbers. Therefore in no state of the atom can two electrons agree in all four quantum numbers" (Ref. 73, p. 123). Quantum numbers were first introduced as indices for the different orbits of electrons around the atomic nucleus, in order to relate the corresponding spectral lines to atomic magnitudes. In about 1924 it was possible to reduce the multiplicity of spectral lines to four indexes of electron motion. Three of them were related to the width, shape, and orientation of the orbit. There was uncertainty about the fourth. Only in 1925 was this quantum number identified as the "spin" of the electron, which was to characterize a sort of self-rotation, a spinning property of the electron. In the discovery of his principle Pauli expressed doubt whether the language and the concepts applied to the movement of atomic electrons, as regularly used by physicists up to that time and still today underlie our easy-to-understand representation for the layman, can describe at all correctly the relationships within the atoms:

> Hence I speak...of an electron in a..."state." I will always avoid the word "orbit"...I believe that energy and momentum values of the stable state are something much more real than "orbits."
> The (still unachieved) goal must be to deduce these and all other physically real, observable properties of the stable state from the (whole) quantum numbers and quantum-theoretical laws. We may not, however, let the atoms become tangled in the chains of our prejudices (including, in my view, even the assumed existence of electron orbits in the sense of ordinary kinematics), but we must rather fit our concepts to experience (Ref. 81, p. 189).

What Pauli asked for in these letters in 1924, Heisenberg translated into reality in the summer of 1925. "Concerning the Quantum-Theoretical Reinterpretation of Kinematic and Mechanical Relationships" was the title of Heisenberg's paper in which he founded matrix mechanics. In this classic text of quantum mechanics Heisenberg developed a method of description with "quantum-theoretical magnitudes," using not the parameters of electron "orbits," but only the frequencies of the spectral lines between different "states" of the electron. These "quantum-theoretical" values obeyed rules of calculation which were known to mathematicians as "matrix algebra." Matrix algebra brought a strong infusion of mathematics into quantum mechanics, permitting statements only in mathematical formulas, along with a sweeping loss of clarity. But the success he achieved in explaining quantum phenomena provided the theoretician (Heisenberg) with his justification.

WAVE MECHANICS

Apart from the development of matrix mechanics, in 1923 a French physicist, Louis de Broglie, wrote a dissertation in which he postulated that matter has a wavelike nature. If electromagnetic radiation, in spite of its demonstrated wave nature, also possessed a particlelike character, could not then matter as well, in addition to corpuscular nature, also exhibit wavelike properties? It is not surprising that Einstein, who had so thoroughly analyzed the question of wave-particle duality in his own works, found this idea interesting and disseminated de Broglie's thesis. The Austrian physicist Erwin Schrödinger, who like Einstein had worked on questions of "gas degeneration" (see p. 69ff), was better suited than almost any other theoretician of the time to take on this question. The adherents of the Munich, Copenhagen, and Göttingen schools were too deeply involved in questions related to spectroscopy to give more than fleeting attention to de Broglie's analogy between light quanta and matter quanta. Schrödinger did not belong to a school and was disposed to consider quantum questions even where there was no radiation involved. He picked up where de Broglie left off and developed the wave hypothesis of matter into wave mechanics, which described the motion of a particle by a wave equation.

With the development of Schrödinger's wave equation there was now available a theoretical approach that was entirely different from matrix mechanics to describe quantum phenomena, an approach that was clearer and simpler to apply and at the same time led to the same results as matrix mechanics. The clarity of the Schrödinger equation was, of course, only on the surface. What, for instance, was meant by the amplitude of a wave of matter? In the case of a vibrating string, amplitude meant simply deflection. In the case of a light wave the amplitude was a measure of the intensity. Both de Broglie and Schrödinger at first thought that the amplitude of a wave of matter was a measure of the density of distribution of mass. If so, was the electron circulating around the atomic nucleus really mass somehow spread around like butter? Max Born in

1926 developed the interpretation that is viewed as correct today: Schrödinger's wave function is a measure of the probability of finding a particle at a given place (Fig. 28).

With this interpretation quantum mechanics took on a fundamentally statistical character. Statements about quantum phenomena became probability statements. A theoretical uncertainty, an indeterminism, had found its way into physics which was different from the long familiar statistical physical laws of heat movement, where the uncertainty stems from the large number of unknown parameters. The inventors of wave mechanics, de Broglie and Schrödinger, as well as the real author of the quantum hypothesis, Einstein, could not accept such a theoretical indeterminism in the course of natural events. "God doesn't throw dice" or "I can feel no reverence for a God who spends all His time playing dice" are two statements by Einstein on the subject. Even with all the controversy over the correct interpretation of quantum mechanics there was agreement among physicists about the correct application and usefulness of the new tool for studying atomic processes. The following sections show how in a short time with this new tool our understanding of the electronic properties of solids took shape.

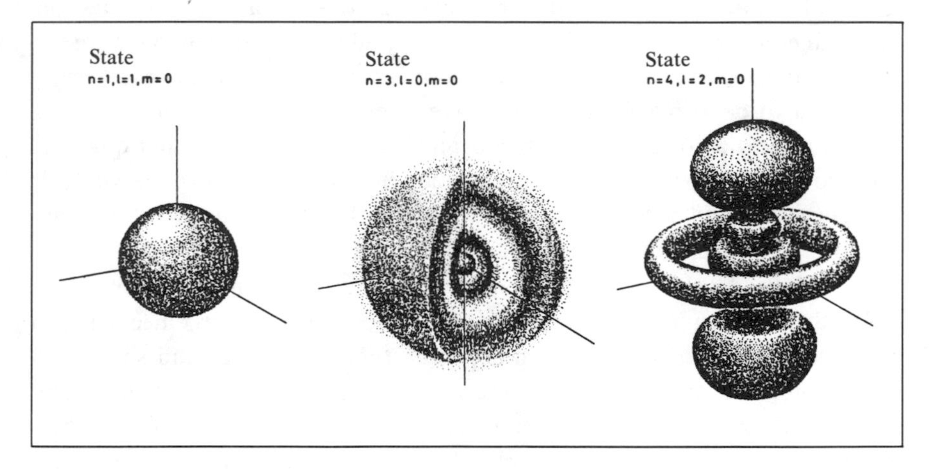

FIG. 28. According to quantum or wave mechanics the location of an electron in the atom can only be characterized by giving probabilities. Shown above is the probability of presence for the electron in the hydrogen atom in the ground state (left) and in various excited states. The blackened regions indicate places which the electron visits in its movement around the atomic nucleus more frequently than the lighter sites. It would be wrong, however, to interpret the spherical or annular shapes as "orbits" of the electron.

SOMMERFELD AND ELECTRONS IN METALS

Our understanding of the electrical properties of matter was based since the beginning of the 20th century on the idea that electrons were present in solids both in the bound state and as freely mobile carriers of an electrical charge. Whether a substance was classed as a metal, a semiconductor, or an insulator depended on the number of free electrons. Theories were developed for metals that described the free electrons as a gas and, as in the classical kinetics of gases, properties such as heat and the current transported by the free electrons could be calculated from the collisions of the gas particles. The concept of the free-electron gas provided the basis for stating that metals are good conductors not only of current but also of heat. The concept resulted in contradictions relative to other phenomena, however. The specific heat of metals, for example, is not noticeably greater than that of insulators. This was inexplicable classically, because heat could be transmitted within the interior of metals not only by the oscillation of atoms but also by the gas movements of the free electrons, hence the specific heats of metals should considerably exceed that of insulators.

After the discovery of quantum mechanics, the dilemma could be explained. The electron gas in metals, because of the small mass of the electrons, is in a completely "degenerate" state already at room temperature, which for gases consisting of atoms or molecules is the case only at very low temperatures. Degenerate means that only a small fraction of the electrons can take part in the heat motion of the cores of the metal atoms. This portion alone (about 1% of all electrons at room temperature) contributes to the specific heat of metals. This theory took into account only the gas-kinetic behavior of electrons in metals as called for by quantum mechanics, but otherwise it followed the pattern of the classical free-electron gas model. Nevertheless, it already could explain qualitatively a large part of the electrical properties of metals. Only after there had been a full quantum-mechanical analysis of electron motion in the lattice of metal atom cores was there an explanation for why electrons are freely mobile everywhere in metals and why other substances are semiconductors or insulators. As we trace the history of these theories, the basic outlines of which appeared in the short space of time between 1927 and 1933, the name Sommerfeld crops up again and again: as architect of the first, still quasiclassical free-electron gas theory in 1927, as the teacher of the people who produced the full quantum-mechanical version, and as propagandist for the new theories both in Germany and abroad. Looking at "Sommerfeld and electrons in metals" as an example, we see that there is more to the development of a new theory than the intellectual struggle for the correct ideas and the formulation in textbooks of the knowledge gained.

THE SOMMERFELD SCHOOL

Heisenberg and Pauli, the stars of quantum mechanics, were not the only celebrities from the Sommerfeld school. The indexes of modern textbooks on solid-state physics—to stay within this one discipline—contain the names of a series of other Sommerfeld students, including Hans Bethe, Peter Debye, Peter Paul Ewald, Fritz London, and Rudolf Peierls. Present-day physics students encounter these names in terms which are part of the vocabulary of modern solid-state physics: Debye temperature, Ewald sphere, London equation, and Peierls umklapp processes. This is not the place to explain these terms; rather here we shall look at the conditions that made this school of physicists so successful.

Certainly Sommerfeld's pedagogical gift played a major role in the success of his school; but there were many able teachers around who were not associated with the establishment of anything that even approached the "nursery of theoretical physics" (as Sommerfeld himself characterized his Munich institute). What were the factors that contributed directly to Sommerfeld's becoming this outstanding personality?

Sommerfeld's own scientific career began shortly before the turn of the century in Göttingen, a mecca for mathematics at that time. Here he met Felix Klein, who was not only a world-famous mathematician but also an influential organizer in the field of science with industrial and political connections. Klein's efforts to orient science toward applications, which had already seen some success in the establishment of the Göttingen Association (see p. 19), now met with much resistance, particularly from the institutes of technology, which were fighting for recognition in the shadow of the universities and claimed applied disciplines as their own special province. Klein had an ulterior motive when in 1899 he arranged the appointment of his assistant Sommerfeld to the Aachen Institute of Technology (Technische Hochschule Aachen). With his student teaching in the Department of Applied Mechanics he was hopeful that he would now have a credible spokesman to explain his efforts to his technologist adversaries.

Because of his central position within organized science, Klein was involved in many activities. One of them was the publication of a multivolume *Encyclopedia of the Mathematical Sciences.* He assigned Sommerfeld to edit the physics volumes. This task consisted basically of finding famous physicists to write overview articles, and it not only served to familiarize Sommerfeld with the broad spectrum of current physics activities at the beginning of the 20th century, but it also caused him to become acquainted with most of the outstanding physicists associated with those areas. When Sommerfeld received an appointment in 1906 to the respected chair of theoretical physics at Munich, it was with the recommendation of such famous physicists as Hendrik Antoon Lorentz, Ludwig Boltzmann, and Willy Wien, with whom he had corresponded for many years on matters related to the encyclopedia.

Sommerfeld's method of working at the beginning of his career was, not surprisingly, completely different from that of other important theoretical physicists of his day, for example, Planck and Einstein. Being less oriented toward fundamental physical questions than toward solving real problems with the mathematical methods so familiar to him, his objective was to involve a large number of gifted students in actual research.

Here, as in his technological work in Aachen, demonstration of the capability of mathematical-physical methods frequently had for him a higher priority than the physical problem being worked on: "Once a problem was broached he was seized, so to speak, with a sportsman's mathematical ambition...and wanted to do the calculations all the way to the end, whether it was important to do so or not" (Ref. 70, p. 533). In contrast to other theoretical physicists, for whom the great flood of students in the "golden age of physics" made it sometimes difficult to find appropriate subjects for doctoral research, Sommerfeld had no difficulty in this respect. He chose topics less for their physical substance than for the applicability of particular mathematical methods. P. P. Ewald reported how he searched in 1908 for a topic for his doctoral research under Sommerfeld:

> ...Sommerfeld took a piece of paper out of a drawer on which were written ten or twelve subjects suitable for doctoral dissertations. They ranged from hydrodynamics to improvements in the calculation of the self-induction of coils, and they also included various problems related to the propagation of waves in wireless telegraphy—every problem listed was a sound exercise in the solution of partial differential equations with special boundary conditions" (Ref. 36, p. 37).

Seminars and colloquia were the appropriate media through which to master the Sommerfeld operating style. A good seminar lecture was a prerequisite if one wished to complete one's doctoral work under Sommerfeld. Outside guests were also invited to the colloquia; for example, in 1909 Max Planck's assistant from Berlin, Max von Laue, attended. Sommerfeld had enlisted him to write an article for the encyclopedia on "wave optics." Thereafter Laue "shifted his lectureship" to Munich and worked as an unsalaried lecturer (privatdozent) until the summer of 1912 in the group around Sommerfeld. All participants in the experiment on "interference phenomena with x rays" in April 1912 came from this group (see p. 59). This experiment, as we have seen, is celebrated in modern textbooks as the "beginning of solid-state physics," because in it the regular arrangement of atoms in crystals was demonstrated.

After the First World War a first generation of physicists had already been reared in Sommerfeld's nursery. These in turn established other important centers of expertise. In the 1920s Debye and Paul Scherrer established at the Federal Institute of Technology [Eidgenössische Technische Hochschule (ETH)] in Zürich, a school that was to become an important center of experimental solid-state physics. Ewald accepted an appointment in 1921 at the Stuttgart Institute of Technology, where he headed a research center for crystal studies until his emigration from Nazi Germany in 1938. The students of Sommerfeld "sold like hotcakes." This is the way Ewald characterized the period around 1920. Sommerfeld's political position in the science of the day was an important reason for this. In 1918, he was elected Chairman of the German Physical Society (as successor to Einstein)—an honorary position which made him a central personality in the German physics community as far as the physics press was concerned. As a member of specialized committees of research support societies, he also had a say in the distribution of research fellowships and assistance in kind.

Sommerfeld's influence was not limited just to Germany. In Kaiser Wilhelm's Germany science, especially physics, was viewed also as an instrument of cultural imperialism.[159] After Germany lost the war, science became the last "bastion of Germany's international standing," as the only remaining asset of German foreign policy, and thus it acquired the status of a power substitute.[43] Since after 1918 German science was subject to an international boycott for several years, Sommerfeld used his influence entirely in this direction. He was offered an appointment as "Representative of Theoretical Physics" for the 1922/23 winter semester under the "Carl Schurz Memorial Professorship" in Madison, Wisconsin. The professorship was founded in 1911 to "invigorate scientific relations between Germany and the United States." "I consider it my duty, in spite of all considerations to the contrary, to accept the appointment," wrote Sommerfeld after the invitation was received. He also accepted invitations to give lectures at the Bureau of Standards in Washington, at Harvard University, and in Pasadena, California. Via his teaching activities in the United States he simultaneously exported his highly valued Munich method:

> I understand "colloquium" to mean a gathering of colleagues and research students where I am not the only speaker. Rather I hope to learn here about the work of colleagues in physics and would like the opportunity to have students to whom I will recommend independent study report the results of their work. This corresponds to the way my seminar operates, which I have been running in Munich successfully now for 16 years (Ref. 9, p. 221).

SOMMERFELD'S THEORY OF "FREE-ELECTRON GAS"

The electrons in metals and associated questions had earlier attracted the interest of Sommerfeld and his students from time to time as one of many interesting physical topics. In 1910 the electron theory of metals was the subject of Debye's lectureship qualification paper. In 1913 Wilhelm Lenz, at the Sommerfeld institute, made a stab at a theory of gas degeneration which, had it succeeded, could have rescued the electron-gas concept which had fallen into ill repute about this time. Karl Ferdinand Herzfeld, who joined Sommerfeld's group after the First World War as his adjunct lecturer, gave special lectures on "The Electron Theory of Metals" and "Kinetic Gas Theory" in the early 1920s. The topic thus was not unknown in the Sommerfeld institute, but it cannot be said that the attention given to it was anything more than sporadic.

The situation changed when quantum mechanics made available a new tool for working on the old problem. A prerequisite for renewal of the electron-gas concept was a correct theory of gas degeneration, and this came along in 1926 with quantum mechanics and the Fermi-Dirac statistics:

> The new statistics grew on the soil of wave mechanics, which was boldly conceived by de Broglie, given a solid foundation by Schrödinger, and recognized to be identical with the (even more solidly established) Heisenberg quantum mechanics; it became dominant in all matters related to atomic physics. According to wave mechanics the electron is to be treated as a state... . This concept of the electron as a state instead of as an individual entity involves a different kind of enumeration... . But there's something else in addition which distinguishes the new statistics from the old: the Pauli principle. This says that...every quantum state may occur only once... . Now if, as Pauli says, each of these quantum states may occur only once in the electron gas, we see immediately a very important consequence ...which is characteristic of the new Fermi-Dirac statistics: the states with the smallest quantum numbers (smallest energy) are fully occupied. There is a shortage of space at low temperatures; because of their large excess numbers the electrons cannot assume the states which they would assume according to classical statistics. They are a "people without a homeland." In an electron gas, because of the small mass of the electron, temperatures up to 10,000 °C must be considered to be "low temperatures." Under all circumstances the electron gas is "degenerate," *i.e.*, restricted by the Pauli principle in its possibilities of motion... . How that works out is not clear at all. But the consequences of this postulate must be thought through (Ref. 185, GS II, pp. 385 and 393ff).

In writing to the publisher of the journal *Die Naturwissenschaften*, where Sommerfeld first publicly discussed this new field with these statements, he added:

> Since the subject is of some general interest, I have drafted the paper in as generally understandable a form as possible for publication in your journal... . Naturally I will report at length in the *Zeitschrift für Physik*, and also in Como at the Volta Congress... . The paper sets forth my claim to put in order using Fermi's new statistical methods the old problem of the galvanic current, the Volta difference, the thermoelectric force, etc. (Ref. 186: Sommerfeld to A. Berliner, August 6, 1927).

The first application of the new statistics to electrons in metals did not, however, come from Sommerfeld, but from his student Pauli, who just a few months after the appearance of the papers by Fermi and Dirac used this tool to explain a special magnetic property of metals. Although the obvious thing to do was to use the same method to till the broad field of metal properties, this was not to Pauli's liking. To him it was primarily a matter of testing the new statistics, and not the establishment of a solid-state theory. Sommerfeld became aware of the subject for the first time in the summer of 1927 during a visit by Pauli, who in the meanwhile had himself become a professor of theoretical physics. Pauli recalled later that he had shown his old teacher the galley proofs of his paper, which was already in press:

> "The next day he [Sommerfeld] said to me that he was very impressed and that [the Fermi-Dirac statistics] should be applied to other areas of metal theory. Since I wasn't keen to do that, he made these additional applications himself" (Ref. 210: Pauli to Rasetti, October 1956).

Sommerfeld, who "could never put anything down on paper before he had lectured about it," [27] used his special lecture in the summer semester of 1927 to test the new subject on a small group of advanced students. At the end of the summer semester Sommerfeld discussed his theory at the Colloquium of Munich Physicists, and in September made a presentation on the topic to the international physics elite in Como, where a "small conference of bigwigs" was held on the occasion of the one-hundredth anniversary of the death of Alessandro Volta (Ref. 186: Sommerfeld to James Franck, July 20, 1926). At the end of his paper he summarized: "We have taken the most primitive ideas of the old theory and worked them over using the rigidly prescribed procedure of the Fermi statistics in a new way. These statistics acquire a solid empirical basis when the results are shown to agree with observational information on metal conduction" (Ref. 185, GS II, p. 421).

Although the theory was able to eliminate a few contradictions of the old electron-gas model, "the agreement with observation" was not as great as indicated here. This was already pointed out in discussions at the Volta Congress. Much additional criticism arrived at the Munich institute by mail starting in the fall of 1927. A respected metallurgist from England (William Hume-Rothery), who was working at that time on a book about the properties of metals, wrote to a colleague: "...I am now wrestling with the new Sommerfeld theory, which is very difficult. I think it is wrong in spite of its success in some quarters" (Ref. 15: Hume-Rothery to W. L. Bragg, January 17, 1929).

At the beginning of the 1927/28 winter semester this widely discussed topic was ready for a basic revision. Electrons in metals became the main topic for seminar papers, and guest fellows from the United States, who had arrived at the Sommerfeld institute with other research plans, were abruptly switched into the new program.

> I went to Munich because I had started studying Sommerfeld's book [*Atomic Structure and Spectral Lines*], and in my mind, he was the authority on all of these things... . When Sommerfeld somewhat discouraged me from studying electron spin he gave me the proof of his first long paper on the application of quantum statistics to free electrons... ." (Ref. 210: Houston interview),

recalled William Houston about his first days with Sommerfeld. (Houston had come to Munich for the winter semester.) Houston was to be concerned with the problem of the "mean free path of electrons between collisions with the atomic cores of metals." Another American (Carl Eckart) was to study emission phenomena, including questions such as: How many electrons are emitted by a hot piece of wire at what temperatures; how does the number of emitted electrons, which are "drawn out" of a metal surface by the application of an electrical voltage, depend on the field strength applied? Still another (Alan C. G. Mitchell) had to busy himself with statistical questions. Bethe, in connection with his doctoral research, worked on the question of the deflection of electron "waves" by atomic cores of metals. A surge of publications issued from this program, and these in turn stimulated further work at other institutions. In the year 1928 alone the *Zeitschrift für Physik* contained 14 papers which were directly related to Sommerfeld's electron theory of metals.

THE QUANTUM-MECHANICAL ELECTRON THEORY

The Sommerfeld theory in its basic form really left unanswered the question of why a free-electron gas existed in metals. Bethe recalled that Sommerfeld "...didn't even care terribly much why the electrons were free, which I thought was a very important thing to know... . I believed the whole theory only when the Bloch paper appeared" (Ref. 210: Bethe interview). In 1928 Felix Bloch was the first to successfully apply the instrumentality of quantum mechanics to the question of the electron states in a crystal. The same problem was posed first for the individual atom and then for combinations of individual atoms in a molecule, and the status of quantum mechanics was substantially strengthened in the process of explicating these questions. Now the regular arrangement of a large number of atoms to make a crystal was on the agenda. How does an electron behave when it is located in the electric field, not of one central atomic nucleus, but of a multiplicity of periodically arranged atomic nuclei? What Bloch and after him Peierls, Bethe, Lothar Nordheim, Alan H. Wilson, and several other theoretical physicists found around the year 1930, was amazing. As a result of the overlapping of the electric fields of neighboring atomic nuclei, the discrete energy state of a single electron which it would have at the site of an isolated atom is split into several closely adjacent states when there are only a few neighboring atoms (as in a molecule), and into a continuous "band" of energy states when there are very many neighbors (as in a crystal)! At the same time the wave function changes. The Schrödinger equation gave, for an electron in a regular lattice, a probability of presence that does *not* (as in the case of a single atom) tend toward zero with increasing distance from the location of an atom, but rather reaches a finite value which again increases with increasing proximity to the neighboring atom, and so on. The electron in a crystal does not, as it were, "hang" from its mother atom but moves with the same probability in the region of other atoms as well. Whether a crystal is an insulator or a metal depends on the extent to which the energy bands of the electrons are filled. A high degree of mobility of electrons in a crystal still does not guarantee high conductivity when the electrons with the extra energy imparted by the application of a voltage can find no free states in the energy band. Crystals with fully occupied energy bands are insulators; crystals with half-filled bands are metals; crystals in which, for example, electron transitions into an otherwise empty band is made possible by the addition of foreign atoms (doping), are semiconductors.

Now it was clear why Sommerfeld's electron theory could be successful after all. For the electrons in an energy band that is not fully occupied, because of the continuous distribution in the band of energy states with external energy changes (voltage, heat, etc.), sufficient empty sites are available in the band so that they behave like a free-electron gas, subject only to quantum statistics. All influences affected by limitations due to the form of the energy bands can be properly explained only within the framework of this new quantum-mechanical theory. We must leave it to the physics texts to provide an insight into the complicated relationships of the electronic properties of solids and their band structure (see the Appendix, p. 59ff).

SPREAD OF A THEORY

Sommerfeld no longer participated substantively in the quantum-mechanical development of electron theory. This he left to his own students and their students. He knew how to help their careers through his influence in the politics of appointments and the distribution of fellowships. Two prime centers for the quantum-mechanical formulaton of electron theory were established at Leipzig and Zürich, where in 1927 and 1928 two exponents of quantum mechanics, Sommerfeld's students Heisenberg and Pauli, were appointed professors of theoretical physics.

In 1926 Leipzig was still a stronghold of old-fashioned classical physics, imprinted with the views and customs of the old-fashioned Otto Wiener and Theodor des Coudres. In the quarter century of their regime practically no penetration of the new quantum physics ideas into the routine activities of the institute was allowed. The sudden death of des Coudres in October 1926 and of Wiener in January 1927 made it necessary to appoint replacements of the professors of theoretical physics and experimental physics. Debye, from the ETH in Zürich was a favorite for the position as head of the Leipzig Department of Experimental Physics, and in June 1927 he wrote to Sommerfeld:

> "If the Leipzig thing works out, I would like best of all to have Heisenberg (in the chair of theoretical physics)" (Ref. 186: Debye to Sommerfeld, June 10, 1927). Sommerfeld suggested to Heisenberg (at the behest of Debye) that he accept the Leipzig professorship, which is especially attractive with Debye and Wentzel [another Sommerfeld student] as colleagues (Ref. 186: Sommerfeld to Heisenberg, June 17, 1927).

"Things worked out at Leipzig" for both Debye and Heisenberg. By the winter semester of 1927/28 the prerequisite staffing had thus been completed for a radical turnabout in physics at Leipzig.

Now a successor had to be found for Debye at Zürich. Here too the matter was settled among Sommerfeld students.

> The Professor of Experimental Physics, Paul Scherrer, who originally also wanted to have Heisenberg, wrote on November 5, 1927, to Debye: "I received a letter today from Heisenberg telling me that he doesn't want to come to Zürich. I am very disappointed, but I am glad that at least you will get him. That way he'll remain in the family, so to speak... . I'd like to ask you something else: what would you think if I offered a position in Zürich to Pauli?" Debye answered on November 26: "He [Pauli] is such an outstanding scientist, and I have always found it so easy to get along with him that physics at Zürich will be well served by this appointment" (Debye papers). Pauli accepted the offer of appointment with effect from April 1, 1928.

The physicists in Leipzig and Zürich formed a group for which comparison with a "family" was not too far off the mark. Debye and Scherrer could look back on collaboration as colleagues and friendship lasting more than 10 years. Heisenberg, Pauli, and Wentzel (who received an extraordinary professorship at Leipzig) had all studied under Sommerfeld at the beginning of the 1920s. Debye's student days with Sommerfeld were 20 years earlier. For each one of them Sommerfeld remained a sort of father figure to which they acknowledged ties even after their own careers were assured: "...best wishes from your old friend W. Pauli (and from Scherrer). Wentzel sends his best and will be writing you very soon. Last evening I sat up drinking schnapps with him until midnight" and "...I hope you continue for a long time running a nursery school for physics babies like you did for Pauli and me [Heisenberg] in our day!." [186]

These are eloquent examples of the way things were between Munich, Zürich, and Leipzig. There was a relationship between Heisenberg and Pauli that reached far beyond their common Sommerfeld "descent." Especially during the 1920s the two shared their separate scientific development more closely than physicist friends had scarcely ever done before. At the same time their characters could hardly have been more different. Heisenberg was open and friendly, as ambitious in sports as he was in science, an early riser, close to nature. Pauli was aggressive and often offensively critical, moody, always preferring a visit to a night club to a morning hiking trip. But, in common they had a fanatical enthusiasm for physics. Here there was a real exchange of their personalities, and when they became heads of their own institutes, an exchange of their scientific thoughts with those of students and assistants.

> Heisenberg wrote once to Pauli: "Do you want to have Peierls as assistant next semester? Of course that is quite all right with me in principle. I feel, however, that you must make it up to me by sending good physicists to L[eipzig]. Can't that be done? I would be very happy if we could set up a sort of exchange of physicists between Zürich and L[eipzig]. But it must be reciprocal; otherwise at the end I would be sitting here alone" (Ref. 81, p. 517).

When Heisenberg turned to Munich in order to recruit assistants for Leipzig from the Sommerfeld school, Sommerfeld answered:

> Your letter reeks of a guilty conscience from beginning to end. So you want to steal assistants? And naturally the best ones at that! My conscience is clearer than yours and I showed your letter to Unsöld. Even if he had a burning desire to go to Leipzig, he still agrees with me that he can't leave now, in the middle of the semester, so to speak (he is also very useful here for the seminar)... .[186]

Although Heisenberg in this case was turned down, a brisk exchange was soon thriving. Bloch, who went to work with Heisenberg at the suggestion of Debye, and Peierls, a Sommerfeld student, shuttled several times back and forth between Leipzig and Zürich. The doctoral theses and inaugural papers with which they began their academic careers under Heisenberg and Pauli became pioneering documents of quantum-mechanical solid-state theory. "I have at this time a rather large enterprise going here in Zürich," wrote Pauli in May 1929 to Sommerfeld, "Mr. Bloch is busy right now working out a theory of superconductivity... . Mr. Peierls is hatching a theory of heat conduction in solids" (Ref. 81, p. 503).

Research fellows were now streaming to the new centers just as they did to the Sommerfeld institute in Munich. The Russian Lev Landau and the Americans Isidor I. Rabi and Julius Robert Oppenheimer, for example, spent 1929 in Zürich. The stream of fellows continuously brought new ideas to both centers and contributed to the rapid dissemination of the latest results. Important international meetings were held. At the ETH in Zürich, for instance, an event took place, which was carried on as "an established institution for several years," an international congress "on experimental and theoretical x-ray physics," which was attended by "prominent specialists from various countries" (Ref. 81, p. 497). In Leipzig Debye institutionalized, starting in 1928, annual lecture weeks, each of which was structured around a topic receiving current research attention and brought together international experts in the relevant field of theoretical and experimental physics. Problems in solid-state physics were prominent topics in each of these "Leipzig weeks." It was the explicit wish of the Ministry of Education of Saxony that these meetings should be international in character, an easily understandable requirement given the role of Leipzig as a fair city with an economic structure directed toward exports.

Sommerfeld's role in the quantum-mechanical electron theory and its dissemination did not consist only in the institutional strongholds established by members of his school in Leipzig and Zürich. He himself exerted considerable influence through his heavy travel schedule as well. In the fall of 1928 he undertook a trip around the world, spending six months in India, China, Japan, and the United States. This trip offered him the welcome opportunity to escape the "threatening" festivities for his sixtieth birthday. Thirty Sommerfeld students, most of whom had become renowned university professors themselves, honored their teacher with a commemorative publication and sent their best wishes, while the head of this band of physicists, which had grown to be a large family, was spreading the word on new developments in theoretical physics through gas lectures in Asia. In this way, for example, Japanese students at the University of

Tokyo heard of the new electron theory for the first time. Sommerfeld's new electron theory also gained attention in industry. Walter Schottky, who was working at the time on the physics of electron tubes for Siemens, corresponded with Sommerfeld on questions related to the work function for electrons emitted by metals. When Sommerfeld made a trip to the United States in 1931, he lectured on the topic during the Sommerfeld Symposium at the University of Michigan. Walter Brattain, who worked at Bell Laboratories as an experimental physicist, and more than a decade later would be one of the inventors of the transistor, was allowed to participate in this seminar by his employer in the hope of gaining theoretical "know-how" for a better understanding of electron tube problems. Stimulated by Sommerfeld's lecture, Brattain subsequently instituted a series of special lectures at Bell Labs to introduce the new knowledge about metal electrons into industrial practice.

Further visible fallout from Sommerfeld's propaganda efforts for the new theory can be seen in the form of review articles written in the early 1930s on this subject. Peierls, the student of Sommerfeld, Heisenberg, and Pauli, summarized the field for the *Errgebnisse der exakten Naturwissenschaften*; Bloch, the Heisenberg-Pauli student, did a summary for the *Handbuch der Radiologie* (*Handbook of Radiology*); and Lothar Nordheim, another Sommerfeld student, made his contribution to the *Lehrbuch der Physik* (*Textbook of Physics*) by Muller-Pouillet, a book used mainly by experimental physicists. Sommerfeld himself also did reviews in collaboration with his American "fellow" Nathaniel H. Frank for *Reviews of Modern Physics* and with his student Bethe for the *Handbuch der Physick* (*Handbook of Physics*).

3

THE BEGINNINGS OF SOLID-STATE PHYSICS

THE "SOMMERFELD-BETHE" ARTICLE on the electron theory of metals appeared in 1933 in a volume of the *Handbuch der Physik* devoted to the broader subject of "structure of coherent matter." In correspondence between physicists this handbook soon became known as the "solid-state volume." Other chapters in the volume had titles like "Dynamic lattice theory of crystals" and "Structure-sensitive properties of crystals." At the beginning of the 1930s the term "solid state" also appeared for the first time in lectures and as a new theme for meetings of physicists. Thus, for example, Heisenberg in the winter semester of 1930/31 gave a lecture in Leipzig entitled "Quantum theory of the solid state;" Paul Scherrer, the experimental physicist at the ETH in Zürich gave his special lecture in the summer semester of 1930 under the title "Physics of the solid aggregate state." Individual specialized areas of the topic had already been the subject of conference papers, lectures, and review articles, but now for the first time the relevant physics information was appearing under the general concept of "solid state," "solid object," "condensed matter," and similar headings as an established subfield of physics. This originated in large measure from the successful applications of quantum mechanics to solid-state problems. Hence the optimistic expectation prevailed among physicists that all questions on the structure and properties of solids could be answered at least in principle. Heisenberg, for example, told his students in his "solid-state lecture" that at the moment much was still not clear, but he hoped that by the end of the lecture more would be cleared up and that one of his students would be motivated by this perhaps to do some research himself. The 1930s were a pioneering time for experimental solid-state research as well, even though in this case there was no comparable event like quantum mechanics for theoretical physics to pull together the various research efforts under the new umbrella topic of solid-state physics.

The unprecedented period of growth in the 1920s made it possible for industry to build up its research laboratories until the effort was interrupted by the worldwide economic crash. In the course of coping with the economic crisis it became clear that enterprises with well-equipped research facilities more easily withstood competition under the difficult economic conditions.

Once the quantum theory began to explain many properties of solids, with obvious benefits for manufacturing, the large industrial organizations began to make their research more "academic" and made working conditions in their laboratories similar to those in the university institutes.

The first two chapters of this book were concerned for the most part with physics in Germany, mainly with quantum mechanics and its initial applications to problems related to the physics of solids. Germany was the leader in the field of theoretical physics until the beginning of the 1930s. This is reflected, for example, in the large number of foreign "fellows" who learned the physics trade in the 1920s with Sommerfeld, Born, Heisenberg, and Pauli. John H. van Vleck, an American theoretician, for example, recalled that only very few physicists in America kept up with the latest developments in quantum theory.

The situation changed completely in the 1930s. Germany's science under National Socialism in broad areas sank to a provincial level, while America became the new scientific great power. This was seen above all in the field of nuclear physics, which developed rapidly in the 1930s.

Hence, in the following sections only marginal mention will be made of Germany. An exception will be the Pohl school at Göttingen, where a center for the study of insulator crystals developed, which gained worldwide recognition in spite of total isolation from current developments in theoretical physics. First of all, however, we should review the political, sociological, and scientific events in the years around 1933 which led to such drastic changes.

1933

EMIGRATION OF SCIENCE

The end of the "golden age" of theoretical physics in Germany followed immediately upon the seizure of power by the National Socialists. The brown terror against all political opposition and everything "non-Aryan" drove thousands to emigrate. Leo Szilard, a Jewish physicist in Berlin, after Hitler was named Chancellor, always kept two packed suitcases in his office at the Kaiser Wilhelm Institute. He emigrated a few days after the Reichstag fire of February 22, 1933. April 1, 1933, was declared "Jewish boycott day." In Berlin the storm troopers occupied the university and mistreated Jewish professors and assistants. On April 7, 1933, the "Law to Restore the Career Civil Service" brought on the "cleansing of the administrative apparatus and institutions of higher learning of all groups out of favor with the regime." In paragraph 3 it read: "Officials not of Aryan descent are to be retired." All university staff were sent questionnaires in which they had to furnish proof of Aryan descent. "A person is considered non-Aryan if he comes from non-Aryan—and especially Jewish—parents or grandparents. This condition is met if one parent or grandparent is not Aryan," it stated (Ref. 11, p. 34).

Bethe was one of the many affected. On April 11, 1933, he wrote to Sommerfeld:

> You probably don't know that my mother is Jewish. Therefore, according to the new civil service law I am of "non-Aryan descent" and hence not fit to be a civil service official of the Deutsches Reich... . One certainly cannot assume that the antisemitism will die down in the forseeable future, nor that the definition of an Aryan will be changed. Anyhow, I have no choice; I must act accordingly and try to find a place that will accommodate me in some other country... . (Ref. 30, p. 141ff).

Many of those authors of the first review papers on solid-state subjects, which we discussed in the preceding chapter, were in a similar situation. Hans Bethe, Felix Bloch, Fritz London, Lothar Nordheim, Rudolf Peierls, and others, whose names are in the indexes of every textbook on solid-state physics, emigrated. How this emigration affected individual universities is made clear in drastic fashion by the example of the mathematics and physics institutes at Göttingen (see Table V). The *Biographical Handbook of German-Speaking Émigrés after 1933* estimates the number of émigrés from the scientific elite alone at 2400–2500, of whom about 1000 were natural scientists and engineers. Although for this group adapting to conditions in the host countries was not as difficult as for those in literature and the arts, for example, emigration usually meant hardship as well for physicists, chemists, mathematicians, and engineers. Only a few, who had already gained international recognition and enjoyed good relations overseas, were received in the host countries with open arms. Bethe, for example, had already made an international name for himself through the "Sommerfeld-Bethe bible," and also could expect glowing letters of recommendation from Sommerfeld. Most of the emigrés found only short-term positions and had to move from country to country several times. The effects of the Depression were also noticeable in science. In the United States in 1933, for example, out of a total of about 27 000 teaching positions on the faculties of around 240 American colleges and universities, more than 2000 were left unfilled for economic reasons. Career prospects for American university graduates were correspondingly poor. Resentment against foreign competitors from over-

INSTITUTE	ORIGINAL STAFF	FIRED
Mathematical Institute		
Executive director:	Richard Courant*	Richard Courant
Directors:	David Hilbert	
	Edmund Landau*	Edmund Landau
	Gustav Herglotz	
	Hermann Weyl	Hermann Weyl
Senior assistant:	Otto Neugebauer	Otto Neugebauer
Regular assistant:	Hans Lewy*	Hans Lewy
Other assistants	Franz Rellich	
	Werner Weber	
	Heinrich Heesch	Heinrich Heesch
	Rudolf Lüneburg	Rudolf Lüneberg
Mathematical Physical Seminar		
Other assistants	Paul Bernays*	Paul Bernays
	Paul Hertz*	Paul Hertz
	Wilhelm Cauer	
	Werner Fenchel*	Werner Fenchel
Other scientific staff:	Herbert Busemann*	Herbert Busemann
	Emmy Noether*	Emmy Noether
First Physics Institute		
Director:	Robert W. Pohl	
Senior assistant:	Rudolf Hilsch	
Regular assistant:	Gerhard Bauer	
Other assistant:	Rudolf Fleischmann	

TABLE V (cont'd).

Second Physics Institute		
Director:	James Franck*	James Franck
Senior assistant:	Hertha Sponer	Hertha Sponer
Regular assistants:	Günther Cario	
	Arthur von Hippel	Arthur von Hippel
Supernumerary assistants:	Heinrich Kuhn*	Heinrich Kuhn
	Werner Kroebel	
Personal assistant to Franck:	Eugene Rabinowitch*	Eugene Rabinowitch
Institute of Theoretical Physics		
Director:	Max Born*	Max Born
Regular assistant:	Walter Heitler*	Walter Heitler
Other assistant:	Lothar Nordheim*	Lothar Nordheim
Other scientific staff:	Martin Stobbe	Martin Stobbe
	Edward Teller*	Edward Teller
	33 Persons	22 Persons

TABLE V. The Nazi purges put an end to more than the mathematics and physics tradition at Göttingen University. In addition to Born and 16 other professors of mathematics and physics, the experimental physicist James Franck and four of his staff were fired. Only Pohl's institute survived the purges without loss of personnel. (An asterisk indicates this person is a Jew.)

seas was not rare. Walter Elsasser, a theoretical physicist who emigrated from Germany to the United States via France, recalls in his autobiography a conversation he had with the physicist and Nobel Prize winner Arthur Holly Compton at the University of Chicago in 1935. When he mentioned in the course of the conversation that he was there looking for a position, Compton's tone of voice changed and he became "very emotional." He said it was "unethical to give positions in the United States to Europeans when at the same time many American students were out of work" (Ref. 34, p. 196). Compton had even sent a foreign fellowship holder by the name of Wilhelmy back to Nazi Germany. Shortly after he arrived he was inducted into the army and died on a forced march. This case finally changed Compton's attitude toward the emigration problem. Antisemitism was a further source of difficulty. A representative of a U.S.-based Emergency Committee for academic émigrés stated: "...a general indifference of the (American) university world and a smug complacency in the face of what has happened to Germany... . Part of this attitude undoubtedly has its roots in a latent antisemitism..." (Ref. 87, p. 241). American Jewish job applicants also encountered this "latent antisemitism." When a Harvard professor in the mid-1930s was recommending his prize student to various acquaintances in other universities and he added that his student was "a tall rangy Texan and neither looks nor acts like a New York Hebrew," he received an answer from the University of North Carolina that: "It is practically impossible for us to appoint a man of Hebrew birth... in a southern institution." At Purdue University no offer could be made because of "antisemitism among the higher administrative officials" (Ref. 111, p. 279).

Nevertheless, out of all the exile countries the most favorable conditions for acceptance, comparatively speaking, were in the United States. In spite of cutbacks during the Great Depression, physical research and teaching expanded significantly in the 1920s and 1930s. The membership of the American Physical Society climbed between 1920 and 1940 from around 1300 to 3700. In an environment of continuously increasing numbers of physics departments the annual number of doctorates awarded rose from about 20 in the year 1920, to barely 100 in 1930 and about 200 in the year 1940 (Table VI). The number of industrial research laboratories rose from almost 300 in 1920 to more than 2200 in 1940. A brisk scientific exchange—financed and organized by powerful private American foundations like the Rockefeller Foundation and the government National Research Council—had in the 1920s already created contacts between European and American researchers, which became very important for the emigrés after 1933. Although at first 37% of the emigrant scientists and engineers sought exile in Great Britain and only 35% in the United States, ultimately 57% found long-term berths in the United States and barely 11% in Great Britain. The reason for this can be seen in the fact that during those years the British university system was basically stagnating. Neither the number of universities (16) nor the number of students and faculty changed significantly in the 1930s.[87,125,201,157]

TABLE VI. Year-by-year increase in the number of Ph.D. degrees in physics awarded by American universities.

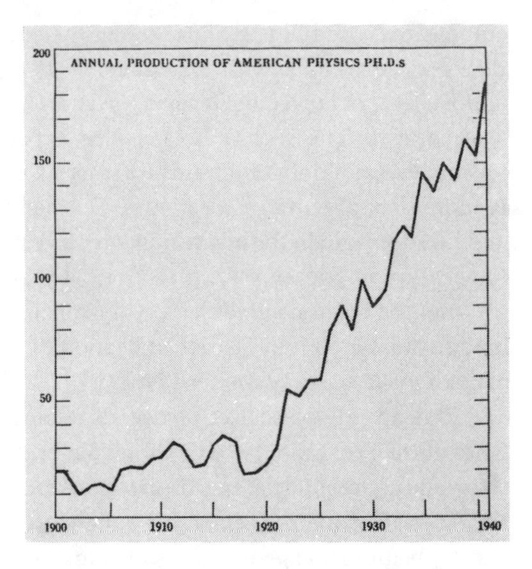

SOLID STATE PHYSICS IN THE SHADOW OF NUCLEAR PHYSICS

The Nazi terror, fascism, antisemitism, and the emigration of scientists that they caused, all hit physicists at a time when dramatic events were taking place in the scientific field. In 1932 James Chadwick, Rutherford's assistant at the Cavendish Laboratory, discovered the neutron, an electrically neutral particle in atomic nuclei unknown up to that time. Harold Urey, a physical chemist at Columbia University in New York, discovered deuterium at about the same time, a hydrogen isotope with double the atomic mass of the ordinary hydrogen atom. Deuterium, whose atomic nucleus consists of one proton and one neutron, represented the simplest test case for the development of a theory of nuclear structure. Also in 1932, Ernest Lawrence and his doctoral student Stanley Livingston at Berkeley succeeded in building a 1-MeV cyclotron (1 million electron volts). This device could be used to accelerate electrically charged particles sufficiently to shatter the atomic nuclei of test substances upon collision with them. These (and other) discoveries opened nuclear physics to both theoretical and experimental investigation. In October 1933 this new subject discipline had already developed sufficiently that a Solvay conference was held on the topic "Structure and Properties of Atomic Nuclei."

Many theoreticians, who had won their first scientific laurels with the application of quantum mechanics to problems of solid-state physics, now devoted themselves to the new challenge of "nuclear physics." In England and the United States, where experimental nuclear physics in particular was strongly emphasized, the theoreticians among the émigrés for the first time came into rather close contact with experiments (in nuclear physics), something that almost never occurred in Germany, where the two lines of endeavor were strictly separated. The separation between experimental physics and theoretical physics was, not only in the areas of nuclear and solid-state physics, traditionally more strongly pronounced in the German university system than elsewhere and had its origin in the institutional requirements prevailing around the turn of the century in Kaiser Wilhelm's Germany (Ref. 45, p. 30ff). Bethe, for example, met the cyclotron-builder Livingston at Cornell University, and the latter made available to him his rich collection of data for the compilation of review articles for *Reviews of Modern Physics.*

But not all physicists turned to nuclear physics. At the Massachusetts Institute of Technology (MIT) John Slater, who in the 1920s had helped shape the quantum revolution in Europe and America, as head of the Department of Physics established a center for solid-state physics. Three of his colleagues shortly before had spent periods of study at Sommerfeld's institute in Munich, working on problems related to the electron theory of metals. Now the theoretical treatment of solid-state problems was entering a new phase. It was sufficiently clear around 1933 that with the aid of quantum mechanics the behavior of electrons in solids could be correctly explained qualitatively; the task now was to describe solid-state phenomena quantitatively, for example, as dependent on the structure of the basic crystal lattice. The program of investigation for the physicists at MIT included such topics as "x rays and crystal structure," "dielectrics and crystals," "magnetism," and "electronics, electrical conduction in gases, thermionic emission, and the photoeffect," for which one or two physicists were responsible. Every spring two- or three-day electronics conferences were held at MIT, with participation by physicists from other universities and industrial laboratories. Bell Laboratories and General Electric often filled their research positions in those days with MIT physicists.

Another solid-state center grew at Princeton around Eugene Wigner, who had received a part-time appointment there in 1930. When the Nazis displaced Wigner from his professorship at Berlin in 1933, he moved permanently to Princeton and became an American citizen. Together with his first doctoral candidate, Frederick Seitz, he discovered in 1933 a simple method for calculating quantitatively the energy bands of metals. The electronic structure of real solids could be calculated by this method for the first time.

BRISTOL

In June 1930 the British government, through the Department of Scientific and Industrial Research (DSIR, see p. 36), approved a research proposal with the title "Theoretical Investigation of the Physical Properties of the Solid State of Matter," submitted by John E. Lennard-Jones, professor of theoretical physics at the University of Bristol. This was the start of the rise of the provincial University of Bristol to one of the leading centers of British solid-state research. The history of this center shows how in this particular case the scientific, economic, and political events of the 1920s and 1930s governed the development of a new independent discipline. The new quantum mechanics, the flood of theoretical physicists from Nazi Germany, and the new challenge of nuclear physics—we encounter all these things again here. Solid-state physics did not develop at Bristol, however, simply from the need of problem-seeking theoreticians for applications of the new quantum mechanics or as the consequence of a sort of "development aid" by émigré physicists, linking their theories developed in Germany with the pragmatically oriented British physics. The Bristol example shows how closely the beginnings of solid-state physics were connected to the increasing need of an industrialized society for better mastery of the properties of materials. For example, at that time the rapidly expanding aircraft industry had a critical need for information on the properties of metal alloys. New materials were available, but there was far from an adequate understanding of their properties on the basis of their atomic composition. Most of the new materials were the result of accidental discoveries or, at best, of systematic testing. At the National Physical Laboratory, the center for metallurgical research in Great Britain, much experience was amassed during the 1920s and 1930s in dealing with a continuous stream of new variations in the composition of alloys, which were of interest mainly for airplane construction. The search for new alloys, however, went on without a knowledge of the physical relationships, such as how the hardness of an alloy was determined by particular additives, or which substances contributed the greatest resistance to corrosion or combined the maximum continuous load-carrying capacity with the desired hardness and temperature properties.

SOLID STATE PHYSICS AT A PROVINCIAL UNIVERSITY

So it is not really surprising that the initiative for support of theoretical solid-state physics at Bristol came, not from the scientists at the university, but from interested groups in industry and government. The key figure behind the new project was Frederick Lindemann, a well-known physicist with extensive connections to the economic and political leaders of Great Britain. He had studied in Berlin before the First World War, and as a student of Walter Nernst became familiar with efforts to provide quantum theoretical explanations for solid-state phenomena. Upon his return to England he went to work during the First World War for the Royal Aircraft Establishment. In the 1920s, a professor of physics at Oxford, he built up the Clarendon Laboratory to be one of the leading research facilities in the field of low-temperature physics. He had close contacts with the powerful industrialists at Morris Motors and the Imperial Chemical Industries. In the Second World War he was also slated to climb to political power as scientific advisor to Churchill. As early as the end of the 1920s he belonged to the advisory staff of the DSIR (Department of Scientific and Industrial Research). It was here that Lindemann in December 1929 suggested that the properties of metals and alloys be the subject of theoretical physical research, to complement the investigations at the National Physical Laboratory.

> The chairman of the committee in the DSIR concerned with this matter, as director of the British Iron and Steel Federation, had wide-reaching influence on the field of metallurgical research. After he had discussed Lindemann's proposal with various metallurgists, he wrote in a memorandum that such studies are "in the national interest" and deserved "all possible support"—"at least until the stage is reached where the possibility of determining, and, as suggested by Professor Lindemann, even forecasting the quantities of different metals can be ascertained with some degree of accuracy." On February 12, 1930, it was agreed at one session that a scientific study on the relationship between the physical properties of metals with their atomic composition should be encouraged. Research proposals on this topic were treated favorably. The theoretical physicist Professor Lennard-Jones of the University of Bristol was discussed as a possible applicant. In 1928 he acted as rapporteur on the Sommerfeld electron theory of metals at a conference of the Physical Society, and he had already himself published relevant papers in this area. Lindemann was to contact the theoretician in Bristol to find out whether Lennard-Jones was willing to undertake research in this field if he were assured of financial support from the DSIR, according to the minutes of the session that day. On March 25, 1930, Lindemann wrote to Lennard-Jones and described to him the "concern" of the DSIR that "in England there was a lack of work on the theoretical aspects of materials in the solid state. The letter stated further: "I am sure any proposals would be very favorably considered as this subject, which is of so much importance to technical problems, is at present almost entirely in the hands of people without that training in theoretical physics, which is essential to its successful investigation".[110]

Lennard-Jones eagerly accepted the offer, whereupon the DSIR approved funds for a research assistant, who should undertake "some theoretical solid-state investigations with particular reference to metals." This position was filled by Harry Jones, a student of the respected theoretical physicist R. H. Fowler from Cambridge, the stronghold of mathematical physics in Great Britain. In 1932 this modest beginning of the "solid-state" program appeared to suffer a setback, when Lennard-Jones left the University of Bristol and accepted an appointment at Cambridge as professor of theoretical chemistry. His successor, Nevill F. Mott, also educated at Cambridge, represented such a good "catch" for the University of Bristol, that Lennard-Jones' departure was more than compensated. Indeed, prior to 1933 Mott had been concerned mainly with current problems in nuclear physics, but he quickly found solid-state physics to his liking. Shortly after his arrival in Bristol the Sommerfeld-Bethe *Handbuch der Physik* article appeared, at just the right time to make the application of quantum mechanics to problems in solid-state physics available to a new generation of physicists.

> I was fascinated to learn that quantum mechanics could be applied to problems of such practical importance as metallic alloys, and it was this as much as anything else that turned my interest to the problems of electrons in solids." Mott recalled later about this period. The head of the physics department at Bristol could be content over the arrival of this barely 27-year-old youth: "no one could possibly have criticised him (Mott) if he had continued his speciality (nuclear physics).... Instead he decided to switch his own interest to metal theory, on which he said his ignorance was profound.... He arrived in Bristol in the autumn of 1933, and within six months he was publishing work in the new field."[110]

The influx of emigrants from Germany beginning in 1933 played an important part in the consolidation of the new center. It was the declared policy of the emigrant assistance organization, the Academic Assistance Council (AAC), to avoid a concentration in the attractive universities like Cambridge and to reduce permanent stays in England to a minimum. These policies were based on an effort to avoid situations in which there would be competition for much sought-after positions at the elite centers, so as not to feed a latent antisemitism and an increasing hostility to foreigners. Still, emigrants with an AAC stipend could at least obtain temporary appointments in the various provincial universities. Walter Heitler, Hans Bethe, Lorenz Frank, Klaus Fuchs, and Herbert Fröhlich spent longer or shorter periods of time at Bristol. They all had fundamental experience in the new quantum mechanics.

THEORIES OF METALS AND INSULATOR CRYSTALS

The program of work for physicists at Bristol differed from that of a German institute for theoretical physics mainly in closer collaboration with the experimental physicists. Relationship to possible applications was an important precondition for the financing of Lennard-Jones' teaching chair by the donors to the University of Bristol, which were mainly the Rockefeller Foundation, the Wills family (tobacco industry), and smaller local contributors. When, following the first successful solid-state calculations by Harry Jones, the significance for technological applications became obvious, and when toward the end of the 1930s additional financial sponsors like the British Electrical and Allied Industries Research Association provided support for physics at Bristol, it was clearly established that no theories divorced from practice would be produced here.

The topics of the studies undertaken can be roughly separated into three areas: (1) The behavior of electrons in metals and metal alloys, (2) the electronic properties of insulator crystals, and (3) mechanical properties of solids. Solid-state studies at Bristol began with problems from area one. In 1934 Harry Jones achieved an important advance in this area when he provided a quantum-mechanical explanation of the so-called Hume-Rothery rules. These are concerned with the relationship between the crystal structure of an alloy and the average concentration of conduction electrons in it. Thus, for example, in the case of brass the crystal structure varies depending on the proportion of the two components of the alloy, copper, and zinc. In pure copper there is one conduction electron for every copper atom. When zinc is added to make an alloy (pure zinc has two conduction electrons per zinc atom), the structure of the crystal lattice changes if the ratio of conduction electrons per atom exceeds certain values. In 1926 the British metallurgist William Hume-Rothery developed empirical rules to determine the electron concentration per alloy atom for which each crystal structure would occur. Harry Jones now succeeded in providing a theoretical explanation for this with the new tool kit containing the band model, Fermi energy, and Brillouin zones (see the Appendix, p. 209ff). In 1936 Mott and Jones included these and other findings in a textbook entitled *Theory of the Properties of Metals and Alloys,* which became the standard work for the first solid-state specialists in the world.[129] This also clearly signaled the shift from the qualitative to the quantitative, a shift which distinguished the first applications of quantum mechanics to solid-state problems (as summarized in the Sommerfeld-Bethe article) from the solid-state theories that appeared in the 1930s. The American solid-state pioneer Frederick Seitz later recalled: "The book by Mott and Jones... opened a new area of speculative work, devoting substantially less effort to the establishment of fundamentals than to the use of approximate models to systematise important areas of empirical knowledge" (Ref. 176, p. 89).

Topics from area 2, theoretical work with insulator crystals, were suggested by a series of experiments at a German center, the Göttingen "Pohl school." The publications from the Göttingen institute discussed the so-called color centers in alkali halide crystals (e.g., table salt NaCl, potassium bromide, etc.). These normally transparent and electrically nonconductive crystals could become colored under certain conditions, such as irradiation with x rays or chemical contamination. The Göttingen experimenters were even able, by introducing electrodes into the crystals and applying a voltage, to create colored "clouds" and cause them to drift through the crystal (see p. 105). These "color-center phenomena" posed an additional challenge for the solid-state theoreticians at Bristol starting in about 1935. Alkali halides at that time were the research field of Klaus Fuchs, who spent four years at Bristol after emigrating from Germany. (Fuchs during World War II belonged to the British contingent at the atomic bomb project in Los Alamos. After the war he became scientific director of the English nuclear research center at Harwell—until 1950 when he was arrested as an "atom spy" and later released to the Soviet Union).

Mott and his research assistant at that time, Ronald W. Gurney, worked out crucial theories in the field of the optical properties of insulator crystals. They published their results in 1940 in a book entitled *Electronic Processes of Ionic Crystals*. The following words appear in the foreword to that book:

> This book was written in order to develop the theory of the movement of electrons in ionic crystals. One of the authors (Mott) had studied for some years the behaviour of electrons in metals, the other (Gurney) the properties of ions in solutions; we were thus able to approach the problems of the conduction of electricity in ionic crystals from somewhat different points of view. We were first attracted to the subject by the very detailed and complete investigations of the properties of alkali halide crystals with color centres that have been carried out in recent years. We have found that an explanation based on quantum mechanics can be given of the great majority of these properties. We found, moreover, that the phenomena observed in alkali halides shed a great deal of light on the more complex behaviour of substances of greater technical importance, much as conductors, photographic emulsions, and luminescent materials.[130]

Another classic of solid-state theory was written in 1940 by Herbert Fröhlich, whose emigration route led him to Bristol in 1935 after a rather long stay in Leningrad. His *Theory of Dielectrics*, which followed his *Electron Theory of Metals* (1936), was already the second book of this Sommerfeld student on solid-state physics.[49,50]

Work on topics in the third category, the mechanical properties of solids, was undertaken at Bristol only in 1939. The important concept for a successful theory of strength and deformation of crystals (see p. 28ff) was the model of "dislocations." Mott once gave the following explanation for this phenomenon.

> Comparison with a carpet fold is very useful... . We all know that there are two methods of pulling a carpet across the hall. One can either grasp one end and pull on it, or one can make a fold at the end and push this carefully to the other end. When a large heavy carpet is involved, the second method requires less effort... . Now we want to look at the situation in a crystal... . What I here call a fold, in a technical language is called a "dislocation"... . It is apparent that the same result is achieved when a dislocation occurs on one side of a crystal and moves through the crystal, as when one half slides over the other... . (Ref. 131, p. 13ff).

In this way the plasticity of materials was related to the movement of dislocations in the crystal structure. The concept of dislocations was still purely hypothetical in the 1930s. Only at the beginning of the 1950s was it possible to demonstrate the physical reality of dislocations. Some researchers at the Cavendish Laboratory at Cambridge, where Mott returned after the Second World War, were once looking at a thin metal foil in the electron microscope, when they noticed a movement. Mott described this discovery:

> I still remember well how they came running to my room: "Professor, come and see a dislocation drift!"... . The possibility of making dislocations in thin foils visible... introduced the study of the whole diversity of alloys used in industry. Every material shows a different behavior; for example, the dislocations in the alloys of copper and aluminum used in building aircraft resemble coil springs, and they actually behave in very similar fashion... . The time when I am writing this (1956) thus in fact marks the beginning of a new field of research, with the objective of explaining the mechanical properties of alloys with the aid of their dislocations, and on the basis of this knowledge to create new materials (Ref. 131, p. 48ff).

Thus solid-state physics, which had such modest beginnings at Bristol in 1930, opened up new territory in important areas of technology, exactly as envisioned by the first financial sponsors at the Department of Scientific and Industrial Research. While most of the work led to practical results in technology only after the Second World War, still its relevance for industry was clear in the 1930s. The work of Mott and Gurney on electronic processes in ionic crystals, for example, led the two physicists in 1938 to a theory about the effect of light on silver bromide crystals (the grains in the sensitive layer of films) and to links with the giant of the photographic industry, Eastman Kodak in Rochester, New York. Contacts with Bell Laboratories also developed after the theories of Mott and Jones on electrical conductivity in metals and alloys had appeared. Fröhlich's work on the theory of dielectrics was supported by the British Electrical Research Association. Mott also maintained continuous contact with the National Physical Laboratory, where the work was concerned mainly with metallurgical problems. A list of participants in an international conference on "The Conduction of Electricity in Solids" held in Bristol in 1937 is also informative.[156] The attendees came not only from domestic and foreign universities (Göttingen, Oxford, Cambridge, and Paris), but also from the National Physical Laboratory and the research laboratories of the General Electric Company in Wembley (England) and the Philips Gloeilampenfabrieken in Eindhoven (Holland). Here, in the Bristol of the 1930s, the alliance of the interests of academic institutions with government and industrial laboratories was already visible in basic outline. This alliance would be institutionalized after the Second World War in many places under the name of solid-state physics and would become a characteristic feature of this new discipline.

THE POHL SCHOOL

When the properties of solids became for the theoretical physicists of the 1920s a test case for quantum mechanics, the researchers turned to the ideal solid, the unblemished crystal. Processes in real solids seemed to depend on so many factors which were difficult to control that their theoretical understanding and the prediction of properties to be expected were very worrisome. "Dirt physics" was the disparaging name applied by some of them to this field, by Pauli, for instance, in a letter to his assistant Peierls (Ref. 123, p. 85). Apart from the circles of the quantum-mechanical elite, there existed since the 1920s a school of experimental physics in Göttingen, which, unnoticed at first by the snobbish theoreticians, amassed a large quantity of data on the properties of solids in complete isolation. For more than 30 years this school was headed by Robert W. Pohl (1884–1977). "The institute was his empire, he did not dominate, but he was in control of everything..." (Ref. 58, p. 6).

The Pohl school, with its investigation of electrical conduction in nonmetals, laid an experimental foundation for semiconductor physics. This group discovered the enormous influence that even the smallest impurities and irregularities in a crystal exerted on its conductivity. This knowledge made it possible later to deliberately influence the conductivity of semiconductors by small additions of foreign atoms (doping).

POHL'S PATH TO GÖTTINGEN

Even as a schoolboy Robert Pohl was interested in scientific hobbies, as is often the case with future physicists. Thus, as a youth he built an x-ray device and worked with wireless telegraphy. Pohl began his study of physics at Heidelberg, because several of his fellow students also enrolled there. But during the second semester he transferred to Berlin. Again it was a friend from his school days in Hamburg who moved him to take this step. As an assistant with Emil Warburg at the Physics Institute of the University of Berlin his friend had promised to arrange a doctoral thesis topic for him early. Thus Pohl began working on an experimental assignment as early as his third semester.

> "I reached that point much too early and hence have often in my younger years confused physics with the building of equipment. I had 'apparatitis,' as they say jokingly in Germany."[153] Pohl received his doctoral degree in 1906 at the age of 21. Decades later (1966) an admiring student wanted to know how Pohl's generation had managed to win a doctoral degree upon barely reaching the age of majority. Pohl wrote back:

> Your question is easy to answer. My friends and I were satisfied with an education which was entirely inadequate and we took advantage of the lax manner in which the examinations were conducted. There was no monitoring of our study which could be taken seriously. Thus, for example, I passed in chemistry with no difficulty even though in my first semester I had only attended lectures in inorganic chemistry and had never taken a laboratory course in the subject! We wasted a very large part of our time as students in hobby-type activities at the Physics Institute... .[154]

In Berlin Pohl became familiar not only with his preference for experimental work but also with the main subject of his future interest, the interaction of light and electrons. This theme runs like a bright thread through the activities of the Pohl school in Göttingen. Pohl studied papers on the photoelectric effect: when light hits the surface of metals, it can release electrons from the metal. It was known that this effect is not restricted to short-wave "hard" light. Visible light as well, and even long-wave infrared light can split off electrons when it hits particular metal surfaces (e.g., alkali metals like sodium and potassium). Pohl also attacked the problem of the photoelectric effect associated with x rays. In the "Berlin Colloquium," where younger participants reviewed the latest publications in physics, he was the "specialist" for the work of Phillip Lenard, who had worked in this field, among others.

Together with Peter Pringsheim of the Imperial Institute of Physics and Technology, Pohl conducted his own investigations on the photoelectric effect, especially on the effects of "variously colored" light. They combined optical and electrical measurements, a method which Pohl also used later to study the conducting characteristics of solids. For their experiments on the surface photoelectric effect the two physicists developed a procedure for producing thin films with clean reflective surfaces by vaporizing metals in a high vacuum. This procedure was further developed at Göttingen to become a standard process for nonmetals as well. Pohl's students learned the procedure and passed it along in turn to their own students. The vacuum vapor deposition technique is being used today on a steadily increasing scale, more than 50 years after its development, mainly to manufacture integrated circuits and for optical films.

During the First World War Pohl was drafted into the service as a senior engineer with the rank of captain in the communications and Zeppelin airship transport areas. The experience he gathered there with electron tubes was of benefit to him after the war in his experiments. On the other hand, because of his military service he could not take up his duties as extraordinary professor at Göttingen until 1919, even though he had been appointed in 1916. When he learned of the offer of appointment, he wrote to his mother, "No one will ever get me to leave Göttingen again" (Ref. 138, p. 20).

His appointment to Göttingen opened to Pohl a sphere of activity in a university with an excellent reputation. In addition to the outstanding staff in the fields of mathematics and mathematical physics, which included Felix Klein, David Hilbert, and Herman Minkowsky, Göttingen possessed a tradition in experimental physics which extended back into the 18th century. Pohl's colleagues Max Born and James Franck, who were offered appointments to Göttingen at almost the same time he was, saw to it that this university also shared in the "golden age" of quantum mechanics. However, while the physicists around Franck and Born worked together as if they belonged to one institute, meeting with their chiefs daily to talk about physics, the group around Pohl was thoroughly isolated.[34] "There were both strong competition and friendly disdain among the three groups. With play on words they called themselves and each other the '*Bornierten*' (in German meaning 'ignoramuses'), the '*Franckierten*' (in German the 'postage-paid'), and the '*Pohlierten*' (in German the 'polished').... And each thought he was the best of all."[139]

The seizure of power by the Nazis signaled an abrupt end to this competition. While the quantum-mechanical physics represented by Born and Franck was driven into exile, Pohl's institute survived the purges unscathed (see Table 5).

THE "POHL CIRCUS"

When Pohl began to build up his group of experimental physicists, the appointment to Göttingen University brought him certain advantages. The university attracted talented students, from whom Pohl could select the best. Pohl had an outstanding mechanic at his side, a man named Sperber, who helped him build his lecture experiments and measuring apparatus. Some of the equipment was copied by the Göttingen firm Spindler and Hoyer and sold commercially. Thanks to a grant from the Rockefeller Foundation the physics institutes could be remodeled and expanded in 1926 under Pohl's direction. In addition to a lavish lecture hall there was room in the main building for an apartment for Pohl on the top floor.

All his life Pohl viewed teaching as one of a Professor's main responsibilities. He felt obliged to follow the example of Georg Christoph Lichtenberg, who as early as the 18th century had made use of the lecture demonstration experiment technique at Göttingen. Lichtenberg's lectures in that day inspired his students so strongly that at times more than 100 listeners filled the benches (out of a total student body of a few hundred).

During his first years at Göttingen Pohl concentrated on revamping the "big physics lecture." He replaced the usual improvised lecture experiments with modular devices that could be assembled for conducting simple and easy-to-understand experiments. A typical Pohl device was the silhouette projection, which could make the experiment clearly visible to a large number of listeners (Fig. 29). It was not just students who showed appreciation with thunderous applause for the experiments which ran like clockwork; the audience often included interested citizens of the city. It was considered fashionable to have attended Pohl's lectures at least a few times. The faultless delivery of the lectures was made easier after the large lecture hall was remodeled. Pohl described his new lecture hall, which several times served as a model for new construction and remodeling elsewhere, in detail:

> ...the gentle rise of the seating gives all of the audience almost the same comfortable angle of view toward the apparatus... . The experimental surface consists of a block of hollow tile 5 m wide, 12 m long, and 30 cm thick... . The surface is not interrupted by any built-in obstructions. Thus it is possible to conduct experiments requiring a lot of room... . Ventilation is an important item. The full stream of fresh air, warmed in winter, flows in from above and below on to the experimental surface and then passes across the audience to the exhaust vents. Thus the instructor is protected from the spent air of the auditorium... . Unfortunately the hall is only 12 m wide. This width was determined absolutely by the structure of the old part of the building. Because of this the number of regular seats was restricted to

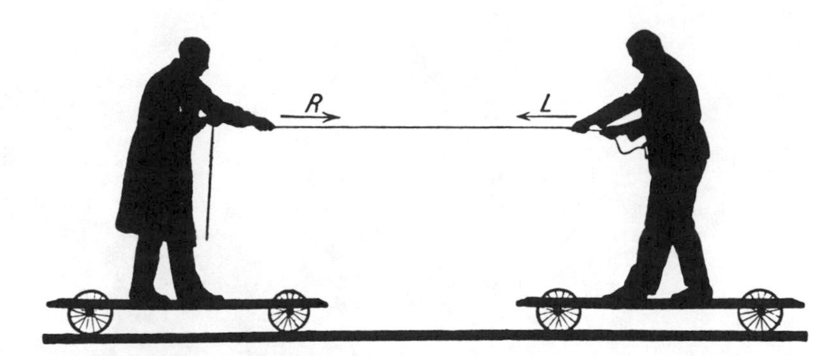

FIG. 29 Silhouette projection to illustrate the principle of force=opposing force. Smaller experiments were also shown to the audience using this method. Pohl is on the right-hand cart and the institute's mechanical specialist Sperber is facing him on the other cart.

400 instead of the 800 which are really needed' (Fig. 30). Pohl had his motto, which he took from Lichtenberg, mounted at the front of the hall. He considered it applicable to both teaching and research: "Simplex sigillum veri" (simplicity is a sign of the truth) (Ref. 126, p. 18).

Pohl became world renowned among his colleagues for his lectures, although critics sometimes characterized them with the words "theater" and "circus." In 1962 he was awarded the Oerstedt Medal by the American Association of Physics Teachers for his original contributions to the teaching of physics. The lectures served as the basis for Pohl's textbooks on experimental physics, the first of which, *Introduction to Electricity*, was published in 1927, and its 21st edition appeared in 1975. Like the volumes on mechanics and optics, it was translated into five languages. During Pohl's 33 years of teaching at the First Physics Institute, 62 students earned doctoral degrees, of whom seven came from outside Germany. Seventeen of them became professors.[126] Pohl built a school of researchers which was unusually strongly attached to him. "He was among the last and most outspoken of authoritarian institute directors. Nothing took place in the institute unless he wished it. Instruction, organization of the work of the institute, of the workshop, of the library, and of the colloquia were systematically planned and executed from the beginning" (Ref. 58, p. 3).

Because of his love of his Göttingen institute Pohl turned down attractive offers of employment from other institutes of technology (Stuttgart, Würzburg, and Heidelberg) and industrial research facilities. The internal structure of the

FIG. 30. View of the large physics lecture hall at Göttingen University. A silhouette projection on the left, Pohl's motto on the right: "Simplex sigillum veri" (simplicity is a mark of the truth). The staff of the institute preferred a less-accurate translation: "Sealing wax is the only truth."

research group led by Pohl remained the same for decades. His assistants (in order: Bernhard Gudden, Rudolf Hilsch, Erich Mollwo, Heinz Pick, and Fritz Stöckmann), after completing their doctoral work, under Pohl's direction worked as senior assistants on their lecturer's dissertations to qualify themselves for independent academic positions. For this it was necessary, "to make certain of Pohl's benevolent support. He enjoyed excellent relations with the Prussian Ministry of Education. They listened to his advice" (Ref. 1, p. 159).

The support of Pohl was almost a guarantee of an offer of appointment to another university. Pohl also corresponded with Arnold Sommerfeld on appointment matters; Sommerfeld also had great influence on the filling of teaching positions (Fig. 31). After Born had left Göttingen University in 1933, Pohl wrote to Sommerfeld:

> You ask about appointing a replacement to Born's position. For the time being nothing can happen there. Born insisted strongly that he be fired, but

FIG. 31. Robert W. Pohl and Arnold Sommerfeld at a meeting of physicists near Braunfels in 1932. Both physicists had a great influence on the scientific profession in Germany. The back of the photo bears the inscription: "A canonical ensemble" (a technical term in theoretical physics).

the Ministry fortunately denied him that, and gave him only a three-year foreign leave of absence... . Hence for the next few years here at Göttingen it will be a matter of an interim appointment. Unfortunately it's clear to us all that the staffing problem won't be easy to solve."[146]

The assistants and the technical staff made up the core staff of Pohl's institute. These were supplemented by students seeking doctoral degrees, and guest researchers from other laboratories, mostly foreign. For example, Abram Fedorovich Joffe in Leningrad, one of the few physicists working in related areas, sent some of his students to Pohl. In March 1928 Pohl answered an inquiry from Joffe as follows: "Of course I'll be happy to receive your Mr. Goldhammer next winter, since the new construction will be finished then. I continue to be quite satisfied with Dr. Arseneva and Mr. Maslakovets; they are both really competent and likeable people."[145]

The staff of the institute numbered at most a total of 10 to 15 persons. The hierarchical structure made it possible to maintain continuity in the research projects in spite of the frequent changes in the investigative staff. More decisive with respect to continuity, however, was Pohl's influence on his students: "scarcely anyone could escape his vibrations." The students adopted the experimental methods of their "respected father figure," his points of emphasis in "seeing the problems," and even his style of writing. "One recognizes the Pohl student after a few minutes of listening to a lecture or after reading a few pages in a textbook" (Ref. 1, p. 1959). The influence was more extensive in the publications of the institute. Almost all papers carry Pohl's name as coauthor (usually in alphabetical order) and if he did not himself write the text, he was involved in producing the final draft.[139] By demanding much, Pohl created a high standard with a scientific atmosphere which motivated his collaborators to work hard late into the night. In order to achieve the necessary accuracy, many measurements had to be repeated 20 times or more. "There was no (upper) limit to the working day, and often Pohl would visit his night workers after dinner."[60]

Pohl also became actively involved in the private affairs of his assistants. During Hilsch's period of military service Pohl wrote some personal lines to his student before bringing him up to date on happenings at the institute:

> Dear Radioman Hilsch! Sincere thanks for your letter of March 26. We obviously were thinking of each other at the same time... . You evidently sat down and wrote to me at the same time. The letter can be very briefly summarized in one sentence: "This is the way things are going." Surely the time will be good for you physically. A complete change like this always turns out to be beneficial.[147]

> One month later he slipped into military jargon when he reported to "Radioman Hilsch" on progress in the diffusion experiments: "I think a general attack on three sides will force the surrender of the fortress."[148]

> He had some equally friendly words in December 1936 for another of his students, "Private Glaser:" "...I have read about your military experiences with great amusement. Things are still the way they used to be, I repeated to myself after reading every sentence. Human nature will never change, and nothing is more conservative than the typical soldier. I ask myself again and again whether it's not possible to make better use of human resources. There is no doubt that war is becoming more and more a technological enterprise, and in assigning people more attention should be paid to the specialized qualifications of the individual. Physically this period will certainly again be very good for you. Your card shows that you are in a good frame of mind."[149]

"THEORIES COME AND GO, FACTS REMAIN"

When, at the end of the 1930s, reports of the work at Pohl's institute were received by theoretical physicists who were also working on the properties of solids, many expressed astonishment that experiments had been conducted there which were important for a theoretical understanding of the phenomena.

> It is a rather remarkable characteristic of this work that, although the investigators have never had a very deep interest in the fundamental interpretation of the properties of the discolored alkali halides in terms of modern atomic theory, the experiments they have carried out have been exactly those that are needed to furnish the basis for such an interpretation. This fact indicates that they possess a very deep intuitive sense concerning the physically important quantities which enter into the problem (Ref. 175, p. 384).

Actually, theoretical interpretations and models play a secondary role in the publications of Pohl's institute. The basis for that, however, was not a lack of interest in a fundamental understanding of the results, but rather Pohl's attitude toward theory. He always regretted his lack of mathematical skill, excused by his extremely experiment-oriented training. Pohl drew a sharp dividing line between experimental and theoretical physics, and viewed as the mission of his institute the gathering of experimental data. He possessed an ambition to acquire empirical data that was so exact that even after 50 years it would have lost none of its currency. Recalling a statement of his teacher Quincke, he often said to his students, "theories come and go; facts remain." He advised them "not to get involved in discussions with theoreticians; we wouldn't understand them anyway" (Ref. 127, p. 5). Theory and experiment should be pursued in parallel and then interact to advance knowledge. When modern theoretical physics (quantum mechanics and relativity theory) fell into disrepute with the advent of National Socialism, in spite of his personal resentment Pohl committed himself to its aims:

> Theoretical physics has been subject to some disparagement in Germany in recent months. Not entirely innocently, of course. Its highly developed artificial language, which is downright unintelligible to the uninitiated, often causes youthful advocates to stray into learned pomposity. Personal opinions are arrogantly considered to be more significant than the facts which do not match them. Empty words and phrases abound. But don't we find lots of hot air in other fields? Must the commendable fight against bombast begin with physics? Aren't there really more promising fields for such zeal? Experimental and theoretical physics cannot be divided and evaluated separately. One can't treasure the music and reject the notes just

> because not everyone can read them. All physical knowledge, which is so very important for technical progress, can be advanced by using only two methods: experiment and mathematical analysis (Ref. 146, p. 7).

Experiments were mathematically analyzed and theoretical models discussed even in Pohl's institute. Very little of this, however, was reflected in the publications of the institute, and what did get published was very carefully formulated. An interpretation was usually followed by the comment that this possibly was not a "tidy explanation." Pohl did not want the publications from his institute to be sullied by speculation.

MODEL MATERIALS FOR SOLID-STATE PHYSICS

Pohl's institute in Göttingen was the first in which experiments on the physics of crystals were conducted almost exclusively after the First World War. His research program, titled "Investigation of Properties of Solids," was ahead of its time and justifies calling his enterprise the first great school of solid-state physics.

> ...gas spectra were fashionable, not solids. There was much discussion of quantum physics in our colloquium. The expression "solid-state physics" was largely unfamiliar, and my colleagues from neighboring institutions were always running to the library to look at the latest issues of foreign journals out of fear that someone on the outside might have anticipated their results. We, however, were quite alone with our ionic crystals and could take our time. Nothing we were doing had anything to do with atomic theory and hence our work was scarcely noticed (Ref. 86, p. 2).

Pohl opened this new field of research when he took up his duties as professor at Göttingen. Following his work on the surface photoelectric effect in Berlin, he turned his attention at Göttingen to the internal photoelectric effect.

> The desire to free ourselves from [the unclear conditions on the surface] caused us three years ago to pursue our study of the photoelectric effect in the interior of crystals rather than on the surface of materials (Ref. 65, p. 332).

A further reason, perhaps, was the fact that Pohl took over a doctoral candidate in Göttingen named Gudden, who had been working for a long time on mineralogical problems and had lots of experience with crystals.[133]

The subjects treated in the 388 publications which appeared between 1919 and 1953 revealed a well-organized research enterprise with centralized long-term planning. The individual papers were mostly small contributions which added a few new empirical facts to the overall area of electronic processes in ionic crystals.

The group succeeded, for example, in generating free electrons by irradiation with light—first in powdered zinc sulfide with admixtures of copper, and later in crystals as well (e.g., diamond, sphalerite, and alkaline-earth phosphors). The freed electrons could be moved through the solid by applying an electrical field. This made it clear that the lack of conductivity in insulators is due to a lack of free electrons and not to some sort of crystal lattice structure which blocks the movement of electrons.

The study of electrons in crystals on the basis of their optical and electrical effects became the most important research program of Pohl's institute. At first only natural crystals were available, and

> ...observations amenable to quantitative evaluation could be expected only when artificially produced crystals of well-defined chemical composition became available (Ref. 151, p. 36).

Attempts to grow crystals began in 1922 with financial support from the Helmholtz Society. Two years later Pohl was able to hire a specialist in crystal growing from a neighboring institute, and he soon could grow alkali halide crystals the size of an apple. Fundamental experiments were conducted on these alkali halide crystals, whose structure is particularly simple, and the results were transferrable to substances of technological importance. Many times the experimenters in Pohl's group pointed out in their publications that these crystals could be viewed as models. Pohl wrote the following about the relationship between the intensity of incident light and electrons split off in semiconductors:

> In Göttingen we have gone to great trouble to answer precisely this question concerning crystalline semiconductors and their light sensitivity. In doing this we initially used substances which were as simple as possible, paying no attention to the technological usefulness of the materials. We used substances which were good insulators and quite transparent (Ref. 152, p. 2ff).

The effect of impurities and defects on electrical conductivity was especially important for semiconductor physics. Even one foreign atom in a million can change the conductivity sharply. The transparency of the crystals was a further advantage in such experiments. Extra atoms and even certain lattice defects in the crystal could trap electrons and in this way produce a bright coloration. For this reason they were called "color centers" (see Fig. 32). Using these it was often possible

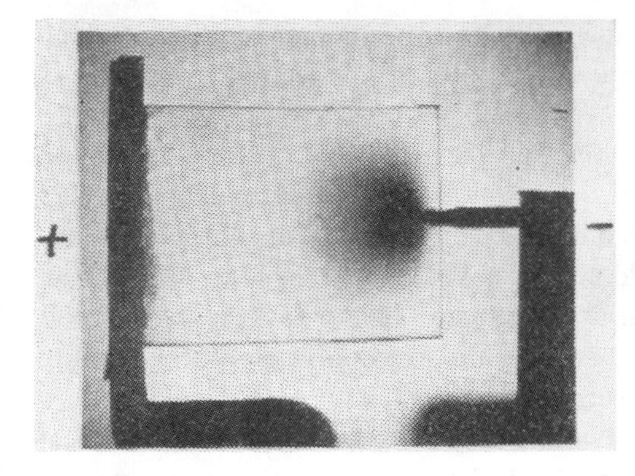

FIG. 32. Color centers in a transparent salt crystal. Distortion sites in the crystal (in the case of vacant sites where halogen ions should really be found) can trap electrons for a time and cause intense coloration. The movement of the electrons in the crystal due to application of a voltage can be observed as the drift of a colored cloud (e.g., blue in the case of potassium bromide).

> to follow electrical processes optically... . Such crystals can be used successfully as models for the optical and electrical behavior of other solids which are not transparent... . The interpretation is now accomplished using a semiconductor model" (Ref. 85, p. 489).

Pohl intended that his models would spur technological development, but he himself pursued no technical objectives. He was rather interested only in the fundamental side of the problem.

POHL AND TECHNOLOGY

An understanding of this fundamental side provided Pohl's students with the best of credentials for finding work in industry, and Pohl said he was always ready to smooth the way for them—to Siemens or AEG, for example.

In 1938 the application of one of Pohl's students for a position as physicist at the Stuttgart Machine-Building School was rejected. Pohl immediately offered his assistance:

> I'm sorry that in applying for the physicist position at the machine-building school you ran up against another example of how capricious human nature can be. I can assure you that no applicant had a better report card to submit than the one which I sent about you to the director... . It's my opinion, however, that you should leave the Stuttgart institute quickly, and if the Bosch firm inquires about you, I'm ready to give them evidence that they'll be running no risk in hiring you. I'd be very happy if you decided to go with Bosch. I have always had a special affection for this plant, although I have had no close dealings with the firm.[150]

When they joined industrial firms, Pohl's students used studies of crystal phospors to build scintillation counters and nonmetallic vapor deposition techniques to reduce reflections from optical components and eyeglasses. They did studies of the photographic process for the film industry in which monocrystals again served as model materials instead of the indistinct granules of the film layers. Pohl's student A. Smakula was hired by the optical firm Zeiss in Jena, where he developed further the technique of growing crystals from a melt with the addition of definite amounts of particular impurities, a technique which had been a prerequisite at the institute for all work with the alkali halides. The technique is used today to produce large monocrystals. Pohl supported practical projects even in fields that were far removed from the institute's primary concerns, such as the rather risky rocket propulsion experiments of his student von Ohain in the courtyard of the university. In January 1941 the Commissioner for Special Propulsion Projects in the Ministry of Aviation reported to the Workshop on Jet Power Plants of the German Academy of Aeronautical Research about

> "...three places [where] individual jet-engine building projects were undertaken earlier without the knowledge and support of the Ministry of Aviation and with very limited resources... . At the Ernst Heinkel Aircraft Plant, the foresight of Mr. Heinkel made it possible for Mr. von Ohain [Pohl's student] to start work on the construction of the jet engine which he had proposed. This effort followed earlier investigations at Göttingen which were conducted with the support of Mr. Pohl, who recommended to Mr. Heinkel that the work be continued" (Ref. 59, p. 183).

One of the most spectacular basic experiments (viewed in hindsight) was undertaken as a result of inquiries from the AEG. In 1938 Pohl and Hilsch published their paper entitled "Control of Electric Currents with a Three-Electrode Crystal and a Model of a Barrier Layer." A potassium bromide crystal was clamped between two different metal electrodes (Fig. 33), and only one of them could release electrons into the crystal. An electric current could flow through the crystal only when the externally applied voltage was such that the electron-releasing electrode acted as cathode. When the polarity was reversed, the electrons remained in the electrode: the crystal blocked them. If a grid was built into the model barrier layer as a third electrode as if in an electron tube, the crystal acted as a current control element. Pohl and Hilsch achieved more than 100-fold amplification of the current with their three-electrode crystal. The experiments, however, did not lead directly to a technical application. Not until 10 years later, with the invention of the transistor in the United States, would it be technically possible to do with a semiconductor device what an electron tube

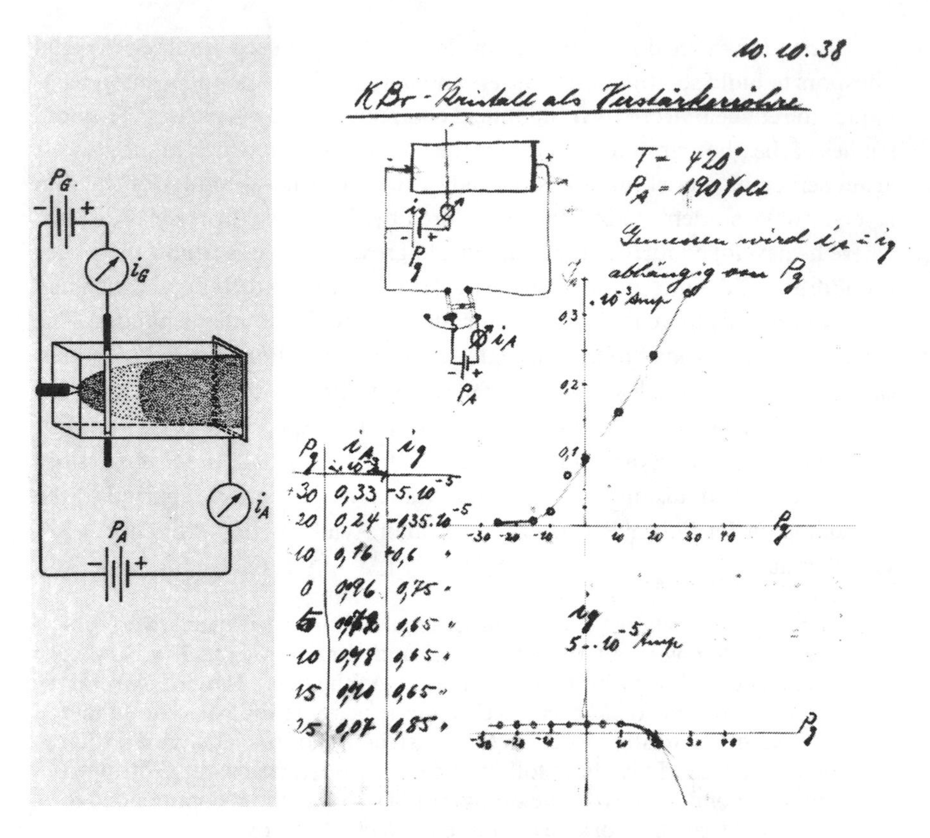

FIG. 33. Three-electrode crystal setup used by Hilsch and Pohl in 1938 to demonstrate that a crystal could be used to amplify current. It functioned like a triode. Because of the low electron drift velocity, however, it was still not usable in technical applications. On the right is an extract from Hilsch's laboratory notebook dated October 10, 1938, which documents the measurements on a three-electrode crystal.

does. Here we see also a typical characteristic of Pohl's model experiments in that they demonstrated the principle of a crystal diode and a crystal amplifier using simple materials. Such crystal amplifiers are not suitable for routine use because of the low drift velocity of the electrons. They offered a large advantage for demonstration purposes, however. The drift of the electrons, and thus their susceptibility to control, could be observed with the naked eye as apparent motion of the color centers.

Pohl was not willing to take the difficult road from physical discovery to economic exploitation. He did not consider it within the responsibility of a university researcher to take out patents. Pohl's students were well aware of his views on this point. Thus, in August 1934 Hilsch turned to his teacher when he invented a simple instrument that could make blind flying easier through use of the forces which appear at rotating objects (Coriolis forces):

> My little model is so good that one could immediately fly off with it. It's lying at my elbow on the desk. I'm sure it could be produced and sold for a couple of marks. But now comes the big question: Do you think, Herr Professor, that a patent besmirches one's character?[83] It appears, however, that Hilsch did not stick to Pohl's principles. One month later he reported: "As far as my patent is concerned, I will indeed work with a lawyer... ."[84]

INDUSTRIAL RESEARCH BECOMES SCIENTIFIC

The centralization and expansion of research capabilities in industry (see p. 20ff) freed the laboratories of commercial organizations from their tight link to production. More and more studies were undertaken that reached beyond direct application and resulted in contributions to fundamental research. When management gave their scientists more freedom, it was not primarily to accede to the demands of their scientific curiosity. Nor did management have any basic interest in fundamental research. At Bell Laboratories the purpose of research was described as follows in 1924:

> The members of the Research Department work constantly in scientific areas which in some way have something to do with electrical communication and accumulate a store of information on which future decisions can be based... . By predicting problems and supplying answers the research and development staff function as an intelligence facility which basically studies the future and supplies the operational and service arms of the firm with accurate information in the technical and industrial areas which they cover. In this way the future objectives can be achieved much more easily and economically than otherwise (Ref. 167, p. 121).

This kind of research, which should bring to light fundamentally new information, but whose goals were not purely scientific, was called "mission oriented." The purpose was "to develop and control any scientific information which (in the case of Bell Laboratories) could be useful for the technology of electrical communication."

UPSWING IN THE 1920s

These ambitious research programs were based on excellently equipped laboratories and contact with the professional scientific community, objectives pursued by increasing numbers of firms in the 1920s. In the United States alone there were more than 1600 industrial research facilities in 1930.

> As competition increased many commercial firms relaxed the early resistance which they had shown toward the hiring of scientists, and instead realized that scientists contributed to their competitive strength and hence to the profitability of their enterprises (Ref. 81, p. 430).

The demand for scientific staff was reflected in recruiting advertisements in the scientific press. In the autumn of 1921 an article in *Scientific American* entitled "Progress in Scientific Research" complained about the dearth of scientists in industry in the following words:

> There is a pressing need for highly trained imaginative workers in industrial research laboratories. It is only during the past 25 years that the research laboratory has been given a place of importance in industry and what development the next 25 years will bring is difficult to foresee (Ref. 206, p. 236).

Positions in industrial laboratories (Fig. 34) still received less respect than academic positions in universities. Industry tried to balance out the lack of prestige with financial rewards. The article cited above had this to say about high salaries:

> There is no definite salary limit for a man of exceptional ability. Charles P. Steinmetz is a research engineer who is paid an enormously high salary and he is one of the foremost scientific investigators of this country. To say the least, the field of industrial and scientific research will give any conscientious worker a good livelihood (Ref. 206, p. 236).

Efforts to recruit scientists for industry also appealed to the patriotism of the researcher. National prestige, an argument at the turn of the century for state support of physical research, now became a motivating factor—in addition to financial inducements—for taking research jobs in industry. In Germany, which lost the war, support of academic research was increased in an attempt to maintain the leadership, which German science enjoyed before the war (see p. 37ff). In America, however, the emphasis was on economic strength.

> "What must America do to establish itself as the leader among nations in making natural forces do the world's everyday physical work?" asked the

FIG. 34. View of an experiment table in the research laboratories of the General Electric Company in 1927.

director of the research laboratory of the electrical giant General Electric at the end of 1921 in *Scientific American*. "We are in competition with others for world trade. The premiership will go to that nation which excels in learning how natural forces may be utilized in the continually expanding tasks of mankind... . We have the...requisites. We have splendid laboratories. We have a wealth of materials with which to work... . But we must have competent men to use them" (Ref. 203, p. 88).

The American universities reacted to the demand with an expansion of their scientific education. The number of scientists available increased, and because of the lack of positions in universities the researchers, who had long opposed the "prostitution of their talents for industry" (Ref. 99, p. 151) had to relax their opposition.

In 1926 more than 600 people were already working in the research laboratories of Bell in fields like electronics, chemistry, metallurgy, magnetism, and crystallography, and an additional 90 graduates of 60 American and European universities were hired that same year. The number of new initiatives rose steadily and had exceeded 350 by 1928–1929.[8]

In Europe also, industrial laboratories became scientific after the First World War. A visitor from the Leningrad Institute of Electrical Engineering, after visiting the laboratories of the largest European firms in 1927, said that he was especially impressed by the Central Laboratory of the Siemens Plant,

> "...which is housed in a six-story building. This Central Laboratory is splendidly equipped as a scientific research institution concerned with all kinds of questions, including those of a purely scientific character, as well as chemistry, physics, and electrical engineering. As soon as...it becomes clear that one or another scientific advance can also be used for industrial purposes, then this is examined thoroughly... . Scientists of international European repute work in all laboratories, and especially those of Siemens and Halske. These scholars are assured indirectly by industry that for them it is entirely unnecessary to hold down several positions" (Ref. 128, p. 1).

Scientific contact with the physics community was maintained through the publications of the industrial laboratories. The first issue of the *Scientific Publications from the Siemens Firm* appeared in 1920, while the *Bell System Technical Journal* was born two years later. In addition to the communication of research results, an important purpose of the latter journal was the education of the Bell Laboratories staff. From 1923 to 1939 a series of articles appeared entitled "Some Contemporary Advances in Physics," in which the newest fields like nuclear physics, atomic physics, cosmic rays, the theory of light, and the quantum theory of solids were discussed. In addition, the scientists were required to publish in the academic professional journals. This was also the way in which the most important research results were disseminated for those laboratories that did not establish a "house periodical" until the 1930s (e.g., *Technical and Scientific Transactions from the Osram Firm* and the *Philips Technical Review*). The number of contributions from a laboratory to the professional literature can be used as a rough measure of the scientific productivity of that laboratory. Based on the number of papers published in the *Physical Review*, one of the most respected physics journals, the Bell Laboratories ranked ninth in 1925 and 1926, behind eight universities led by the California Institute of Technology, Princeton, and Harvard.

In order to familiarize their scientists with the latest developments in academic research, leading firms organized regular educational events. In the weekly colloquium of Bell Laboratories, which was started in 1919, the staff reviewed the latest literature for purposes of discussion with their colleagues. After the first few years more and more external visitors were invited to participate, including famous physicists from all over the world. Arnold Sommerfeld from Munich reported in 1923 on the structure of atoms and in 1929 on the photoelectric effect on individual metallic atoms, Ernest Rutherford of London in 1924 on new studies of the atomic nucleus, Erwin Schrödinger of Zürich in 1927 on his development of wave mechanics, Eugene Wigner of Berlin and Princeton in 1932 on the application of quantum mechanics in chemistry, and Paul Ewald of Stuttgart in 1936 on crystal growth and ordering.[89]

On the other hand, industrial physicists were frequently included in programs of the academic community. For example, in the late 1920s and early 1930s MIT organized a series of colloquia on the applicability of fundamental research to engineering problems. Typical presentations included two by the Bell Laboratories staff on "Vacuum Tube Voltage Amplifiers for Telephony" and "New Research on Magnetic Alloys." Speakers came from such firms as the Westinghouse Electric and Manufacturing Company, the Radio Corporation of America, the Electric Storage Battery Company, Stone and Webster, and the Edison Electric Illuminating Company of Boston. The industrial laboratories also sent representatives to meetings and symposia of the physical societies.

In 1919 a second technical society of physicists (The German Society for Technical Physics) was established in addition to the existing Physical Society. The principal focus of the new society was the applications of physics in technology:

> The purpose of the Society of German Technical Physicists would first be to provide its members with the opportunity, through evening lecture meetings, to share information, to counteract the effects of excessive specialization, to engage in discussion and debate, and finally to make personal contacts... . A further responsibility of the society would be to work to assure the proper training of physicists, namely, in the institutes of technology and the facilities of special physics divisions of those institutes, corresponding to the chemistry divisions... . (Ref. 208, p. 4ff).

Charter members included Gerdien, director of the Siemens Research Laboratory; G. Gehlhoff, director of the Osram Society; Emil Warburg, president of the Imperial Institute of Physics and Technology (PTR); F. Kurlbaum, professor at the Technical University of Berlin; and Karl Wilhelm Knoblauch, professor at the Munich Institute of Technology (Technische Hochschule München).

The flow of information from research in the institutes of technology to industry through lectures and symposia turned out to be inadequate. New concepts were seldom presented in sufficient detail to permit the industrial researchers to apply them to their special problems. The best solution to this problem seemed to be for industry to conduct its own program of fundamental research. Someone with personal research experience in a field should also be able to integrate and adapt the research results of others in that field.

Closer contacts with the universities were needed to accomplish this. For this reason, since the early 1920s advanced students in the United States have had the opportunity to complete their laboratory requirements for graduation in industrial laboratories, and are allowed to attend lectures during working hours. The first student to complete his studies with an appointment to the Bell Laboratories was Lester Germer. He was appointed research assistant in 1919 under the supervision of Clinton J. Davisson, a former professor of physics at the Carnegie Institute of Technology. Their first joint investigation was generated by a patent dispute, which had been going on for years, with the Bell Laboratories competitor General Electric. The dispute concerned the development of thermionic vacuum tubes for electronic amplifiers. The subject matter of the dispute was increasing the emission of electrons through the use of a new type of metal filament covered with an oxide layer.

In studying the spatial distribution of the electrons emitted, Davisson and Germer discovered in 1925 an unexpected pattern that could not be explained by the classical theory. Independently of these experiments Louis de Broglie in Paris had concluded from the fundamental assumptions of quantum physics that under certain conditions electrons behave like waves (see p. 67). In his doctoral dissertation he suggested that the scattering of electrons by crystals be studied as an experimental test of his hypothesis. On a vacation trip to England, Davisson attended a meeting of the British Association for the Advancement of Science, where he became acquainted with de Broglie's theory. He felt that he could carry out the suggested experiments using his equipment at Bell Laboratories. In fact, Davisson and Germer were able to prove in 1927 that electrons are diffracted by crystals like waves. Ten years later Davisson received the Nobel Prize for this discovery.

CONSEQUENCES OF THE WORLDWIDE ECONOMIC DEPRESSION

Both academic and industrial research in physics expanded at the beginning of the 1920s, and the social standing of physicists rose. Increasing business difficulties related to the world economic depression in the 1930s caused enterprises to cut back costs, and this affected the research staff as well. This is confirmed by a sorrowful letter of Gehlhoff, first president of the German Society for Technical Physics, co-editor of the *Zeitschrift für technische Physik*, and director of the Osram Society, to Dr. Sommerfeld. Gehlhoff pointed out to the influential professor from Munich that under existing economic conditions

"...one cannot count on industry's being able to hire the same number of young physicists during the next few years as in the past, and hence the number of physicists in training should be slowed down somewhat in advance." He suggested the adoption of qualifying examination for physicists like those adopted by the universities in the past for chemists at the instigation of industry" (Ref. 187, Gehlhoff to Sommerfeld, December 3, 1930).

The Depression did not affect only the prospects for being hired by industry; researchers with permanent appointments also had to accept penalties. The Bell Telephone Company reduced the work week in 1932 from 5½ to 4 days. The constant threat of losing their jobs motivated scientists to improve their competitive positions. They used the additional free time to improve their knowledge of physics either privately or with university courses. Bell scientists organized themselves into small groups and studied up-to-date textbooks. Alan Holden, a specialist in chemistry and crystals, later recalls this period:

"...we worked terribly hard. We met once a week, after we had worked through a chapter and completed all the exercises, to correct each other's work."[93]

Senior people from the research laboratories of industry began to study again like students. After the ever more successful applications of quantum mechanics to solid-state problems, their relevance for technical applications like electron emission from heated cathodes, photoelectricity, and conduction became clear. Researchers who worked in these fields now needed more knowledge of quantum mechanics. Mastery of quantum mechanics was very difficult for the scientists who were mostly trained only in classical physics. Unemployment and curtailed working hours during the Depression caused many to make the effort anyhow, and these factors thereby updated the knowledge base for practical physics in industrial laboratories.

"FUNDAMENTAL RESEARCH" AT BELL LABORATORIES

At Bell Laboratories, whenever technological problems appeared insoluble without scientific methods, a new research area was always established. As in the case of the search for a telephone repeater in 1911, a new challenge was recognized at the beginning of the 1930s. The electron tube, with its rectifying and amplification characteristics, was still an indispensable instrument for extending telephone communication over long distances, but it was so large, heavy, and fragile. There was an abiding hope that crystals could be made to do the job of electron tubes when many processes within solids came to be understood with the help of quantum physics.

This was not the first time that the electrical industry was interested in solid-state components. Even before 1920 substances like selenium were used in electronics as rectifiers for wireless telegraphy or as photocells for measuring light intensities (Ref. 205, p. 1116ff). In 1925 the Union Switch and Signal Company developed a new rectifier consisting of a piece of copper and a layer of copper oxide. But the physics associated with this "cuprous oxide rectifier" was not understood. As soon as this new component had reached the market, the basic principles governing its functioning were studied experimentally and theoretically at the Siemens laboratory. The main force behind this effort was Walter Schottky (1886–1976), a doctoral candidate under Planck, who worked at Siemens during the First World War but who in 1919 gave up the scientific directorship of the Communications Laboratory in favor of an academic career. As occupant of a teaching chair, however, Schottky kept up his close contacts with Siemens and in 1927 he rejoined the firm full time as scientific consultant. The lack of a functional theory for rectifiers after development of the copper oxide rectifier was described by Schottky later "as not only a technical disadvantage but as one of the largest gaps in understanding in the world of physical phenomena."[170] Schottky and his colleague Deutschmann were able to demonstrate experimentally in 1929 that the rectification processes took place in a thin layer with a maximum of 100 atomic sites. In his "barrier layer theory" of 1938 Schottky was concerned generally with the conduction mechanisms of a semiconductor-metal contact zone. Subsequently an "impoverished layer" appears in the marginal zone of the semiconductor, where only a few electrons are available to carry a current. Depending on the polarity, this impoverished layer is either shifted still deeper into the semiconductor (blocked direction) or completely choked up (flow direction).

Bell researchers also had garnered practical experience with solid-state components. A physicist from Bell Laboratories had become acquainted with the model of the cuprous oxide rectifier at a meeting of the American Physical Society and conjectured that it could be of use to the firm. The material was studied in the laboratory and used for various purposes: for "varistors," components with variable resistance (the resistance depending on the applied voltage and the amount of current flowing); for temperature-dependent resistors, the so-called "thermistors" (Fig. 35), which were used to control amplifier circuits and in aircraft navigation systems. The quality of the components depended strongly on the raw material used. For instance, cuprous oxide elements made

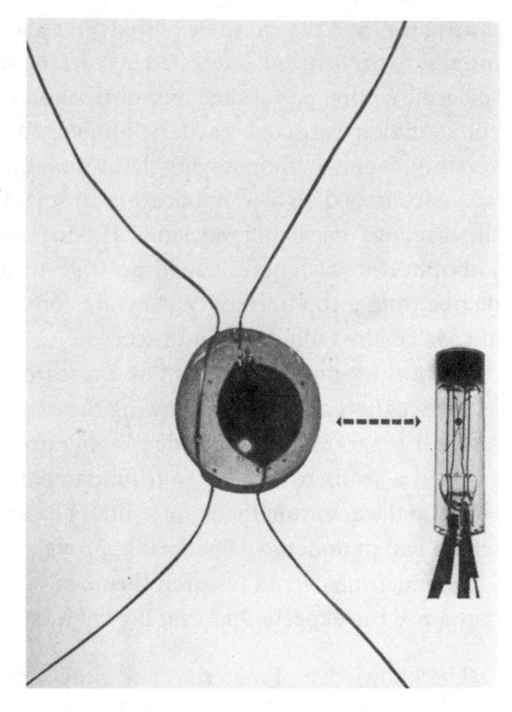

FIG. 35. "The eye that never closes" was how Bell Laboratories promoted thermistors in full-page advertisements (this one in Scientific American). They were able to indicate temperature changes of less than a millionth of a degree and hence could detect the body heat of a person at a distance of a few hundred meters.

of certain types of Chilean copper showed no rectifying effect at all, and there was no recognizable basis for this.

The problem should have been laid to rest by a study of the conduction processes taking place inside the crystals. But not until 1936, when AT&T lifted the hiring freeze imposed during the depression, was it possible to recruit theoreticians with the necessary knowledge of quantum mechanics.

In the same year Mervin Kelly, who had been director of the Vacuum Tube Division, was promoted to the position of research director of Bell Laboratories. Kelly followed the practice of his predecessors and looked around for suitable academic proponents of the new research trend. One of the first scientists to be recruited was Dean Woolridge of the California Institute of Technology (CalTech); another was William Shockley from MIT, who had worked in Slater's group. In the course of his doctoral work he had calculated the diffusion of electrons in sodium chloride, and hence was qualified to study the conductivity of crystals.

In order to attract scientists of this caliber, Bell Laboratories had to create an atmosphere that could withstand comparison with leading universities. The most important attribute was "freedom" of research. Actually, the research was always restricted, because Bell hired only scientists who were trained in fields of interest to the firm and interested in working further in those fields. As far as the scientific equipment of the laboratories and opportunities for interdisciplinary collaboration were concerned, Bell Laboratories were ahead of comparable universities. It goes without saying that contact with scientists in other laboratories was encouraged, as was publication in scientific journals and participation in the meetings of scientific societies. Having spent time as a researcher with Bell Laboratories was equivalent in prestige to having qualified for an advanced degree from a top university. Anyone coming from Bell Labs was and would always be hired almost sight unseen.

Shockley not only worked as a researcher in the laboratory, he also gave lectures to his colleagues. Following the example of the Journal Club of MIT, a seminar program covering wide-ranging up-to-the-minute topics, Shockley organized a seminar devoted to a fundamental and detailed discussion of new work on the quantum theory of solids. The annual work plan of the laboratory, which had to undergo a financial approval process, was submitted for the year 1936 under the overall research theme of "Electrical Conduction in Solids." It contained the expectations that lay behind such "free" research:

> The knowledge of materials and processes acquired will ultimately be of value in the solution of engineering problems. There is also some speculative value in the chance discovery or invention of new materials or processes (Ref. 90, p. 48).

In 1938 Woolridge, Shockley, and Foster Nix were assigned to a research task entitled "Solid State Physics." During the following years the group concentrated its attention on precisely the task which Kelly had conceived when he reorganized the research program: the development of a theory of solid-state phenomena which were of importance in building components for communications technology.

The various research groups at Bell Labs kept in close touch with one another. They could count on the fact that advice was available in a nearby laboratory on many difficult problems, and as a result of mutual stimulation it often happened that several groups were working on separate objectives in very closely allied subject fields. At the request of several colleagues, Shockley and his group concerned themselves with the physics of the copper oxide rectifier. Shockley had just been contemplating the analogy between vacuum tube amplifiers and crystal amplifiers. In the tube the electrons fly unimpeded through the vacuum and can be affected en route by the variable electric field created by a control grid. Shockley now wanted to create the same mechanism in a cuprous oxide crystal. Bell physicists attempted to install fine wires in the copper oxide crystal as control grids in order to affect the flow of electrons. The attempts failed. The newly published "Barrier Layer Theory" by Siemens researcher Schottky suggested to Shockley that he could influence the electrons within the thin surface layer by means of an external field. At the end of 1939 he described the construction of a device which today is available as the "field-effect transistor." It was not possible at the time to verify the correctness of this idea because the quality of the crystal surface was deficient. Shockley interrupted his investigations and left Bell Labs in order to undertake the development for the U. S. Navy of new theoretical concepts to optimize defenses against German submarines.[160]

With the start of World War II, Bell researchers were assigned new priorities. The short-deadline wartime assignments left no room for fundamental solid-state research.

"Industry's basic theme...

...demands the creation of a useful idea and its reproduction thousands of times, until it benefits everyone."

A word from Henry Ford.

When Frederick the Great introduced the mortgage, he really only wanted to help those in Silesia who were wounded in the Seven Years' War to obtain credit. But we see: a useful idea often cannot be stopped and "reproduces itself thousands of times."

Mortages and Community Bonds

The most popular German commercial paper—high interest—priced from 100 DM up at all banks and savings institutions Documented [LOGO] Security

4

WAR OF THE PHYSICISTS

SCIENCE AND THE MILITARY: A RELATIONSHIP WITH A TRADITION

IN EVERY AGE SCIENCE has been used for military purposes. Archimedes is said to have designed a giant catapult to sink Roman warships. Fifteen hundred years later Leonardo da Vinci offered to build for the Prince of Milan any instruments of war he might want—military bridges, mortars, mines, chariots, catapults, and "other machines of astounding effectiveness which are not in general use" (Ref. 209, p. 5). Also, was it not the mission of that seedbed of modern natural science, the École Polytechnique in Paris (see p. 4ff), to build a well-trained cadre of engineers for Napoleon's army? The alliance between the military and the scholarly community also often determined the form and content of scientific endeavor. From the Middle Ages to modern times things like the calculation of the flight path of a cannonball have challenged scientists (Figs. 36–38). This effort ultimately gave birth to an entire discipline, called

FIG. 36. It is clear in this woodcut from the year 1561 that ancient philosophy was not an adequate basis for militarily useful ballistics. According to Aristotle the sphere ejected from the barrel of a cannon follows a straight path until the movement "impressed" on it weakens and it can seek its "natural" location, the center of the world, vertically downward (Ref. 192, p. 129ff).

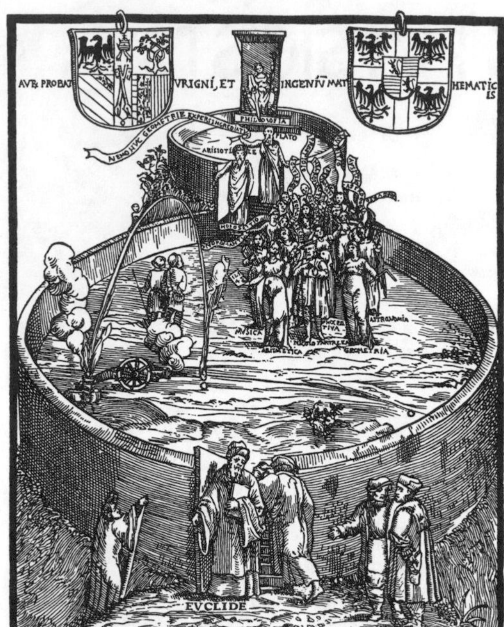

FIG. 37. How does a cannonball fly? This military problem is the subject of debates in the garden of mathematical science as shown in a woodcut from the year 1537. Euclid dominates the entrance with his geometry; Plato and Aristotle give their philosophical blessing to the scene.

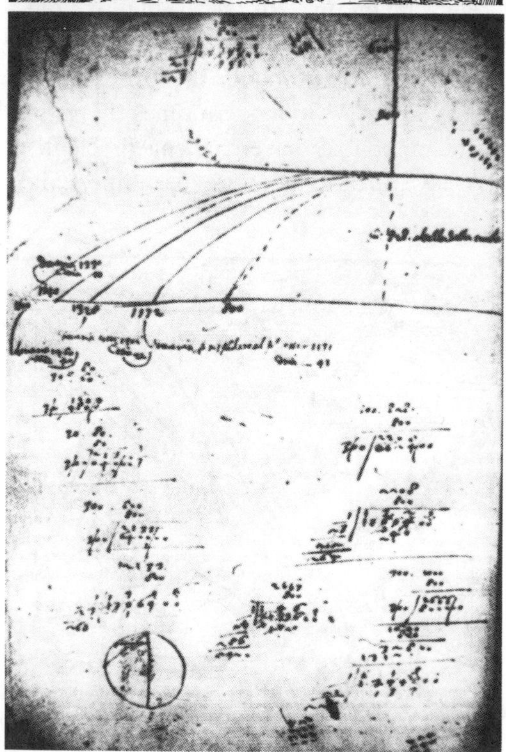

FIG. 38. A Galileo manuscript in which experimentally measured values are compared with calculated ballistic ranges. This manuscript is an example of the eruption of "new science," which proved useful for military ballistics, among other things (Ref. 192, p. 143).

"ballistics," and research facilities, staff, and institutions were dedicated to this field. The history of science is full of examples where military requirements launched new lines of scientific investigation.

Nevertheless, until the 20th century science played only a secondary role in the business of war. The first example of large-scale mobilization of science for military purposes was the enlistment of chemistry in World War I. This led some authors to give the name "war of chemistry" to this first "scientific" slaughter of nations (e.g., Rose, in Ref. 165). Without the large-scale use of chemistry the German army would have run out of explosives and ammunition a few months after the outbreak of hostilities, since the saltpeter needed for producing munitions could no longer be imported. The chemists found a way out of this difficulty. A few years before the war Fritz Haber succeeded in synthesizing ammonia out of air and water on a laboratory scale. In 1913 Carl Bosch proved this method on a large industrial scale. At first the new process, called the Haber–Bosch process after the inventors, was used to produce fertilizer. Ammonia, however, could be further treated to make saltpeter. When a saltpeter shortage appeared after the war started, Carl Bosch made his famous saltpeter promise to the army high command: the Baden Aniline and Soda Factory (BASF), which he headed, could immediately start producing salt-

FIG. 39. Gas attack. "Gas is sprayed from steel canisters and moves in thick billowing clouds toward the enemy trenches and beyond. Protected by this cloud of gas, attacking troops can approach the enemy's position undetected. The gas itself has an additional effect in that it puts out of action an opponent who has not taken precautions against breathing these poisonous and stinking vapors or forces him to retreat. These gases attack the breathing passages and severely damage the lungs..." (Ref. 135, p. 274).

peter for explosives on a large scale from synthetic ammonia, in order to permit the war to continue beyond the year 1915.

When the war on the western front became trench warfare, Fritz Haber, who was acting as a scientific adviser on chemical warfare matters, was busy with a new method of driving enemy soldiers from their protective trenches: poison gas (see Figs. 39 and 40). Haber was subsequently appointed director of the Chemical Warfare Division of the Prussian Ministry of War. In his capacity as director of the Kaiser Wilhelm Institute for Physical Chemistry in Berlin, Haber also had available adequate resources for advancing chemistry as a weapon of war. More than a thousand chemists and technical people worked there on producing and testing suitable poison gases. Other chemical institutes were assigned the task of developing gas masks to permit friendly troops to advance through contaminated territory.[80,69]

[Translation from a German book after World War I. The Poison Gas Weapon, by Professor F. P. Kerschbaum. The development of the German poison gas weapon and gas battle tactics was governed by three considerations: (1) by the effort to confuse the enemy through surprise with new and changing weapons and types of combat; (2) by the need first to dislodge the enemy temporarily from his fortified positions with irritating materials and then, as the forms of combat became more intense, to render him incapable of fighting for longer periods or permanently through the effect of poisonous substances, and (3) by the need to render ineffective the gas protection measures adopted by the enemy. As the course of events unfolds, the compelling logic of the growth of this field is revealed, the interactions between weapons and protective measures become clear, and the close relationships between types of weapons are shown. Before the war and during the war outside the trenches, no one in Germany had any thought of waging battle with substances which have a chemical effect on the human organism. Trust in the unsurpassable fearful effect of artillery weapons was so strong that any thought of using chemicals to weaken or even destroy the enemy seemed unnecessary to the military authorities. But after just a few months it became necessary to reevaluate the effectiveness of artillery in the battle against a fortified enemy.]

FIG. 40. With academic erudition and authority, gas warfare is represented in popular fashion as necessary. The text shown above was taken from a book entitled Technik im Weltkrieg (Technology in the World War), the editor of which wrote in the preface: "Now [in February 1920!] we must make sure that the memory of the victories of the fighting armies as well as the accomplishment of the home guard be kept alive as a bright source of hope which can restore the strength of the tattered, weary, sick, and powerless people and lead them out of the valley of darkness... . Thus let this book go forth as a song of triumph of the bodily and spiritual energy born of the German people, which was never expected to be this strong..." (Ref. 171, pp. III/IV and 278).

Scientists and engineers in other fields also made their knowledge and capabilities available for war service. Physicists, for instance, were used in sound ranging (Fig. 41) or in the operation of x-ray apparatus (Figs. 42 and 43).

On February 4, 1896, only a few months after the discovery of x rays, the *Münchener Medizinische Wochenschrift* (*Munich Medical Weekly*) reported:

> According to the *Reichsanzeiger (Official Gazette)*, the Ministry of War in Berlin has accepted a recommendation to undertake tests in collaboration with the Imperial Institute of Physics and Technology (PTR) to determine whether Röntgen's invention should be made available for surgical purposes in wartime and whether it can benefit sick and wounded soldiers. As a result, a series of photographs of surgical anatomical preparations has been made showing projectiles and parts of projectiles imbedded in soft tissues and bones. The photographs gave a clear picture of the bone damage which occurred and made it possible to determine the site of the imbedded projectile with certainty. The tests were continued on a larger scale (Ref. 61, p. 203).

Fluoroscopic pictures of the skull were presented during World War I as examples of services rended by science.

FIG. 41. The French physicist Jean Perrin in World War I with an acoustic locating device for early detection of approaching aircraft.

FIG. 42. The first use of x rays in war. "In 1896 the medical division of the British War Ministry ordered that two x-ray machines be sent with the expedition [to the Nile] for use by the surgeons to locate bullets in the body and to determine the extent of bone fractures" (Ref. 61, p. 203).

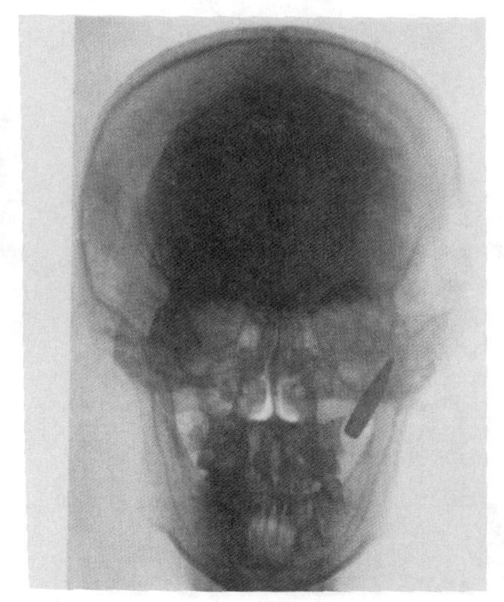

FIG. 43. Example of a skull x-ray photograph from World War I.

> Being flexible and adaptable, as our technology has always been, as well as creative and bold as only science in its diversity can be, natural science, technology, medicine, and agriculture knew how to adapt to the [wartime] situation... (Ref. 168, p. VII).

The large-scale use of chemical weapons in the First World War was possible because chemistry had outgrown the academic laboratory stage in the 19th century, and even before the war could rely on a sizeable industrial base.

With few exceptions scholars, regardless of their discipline, distinguished themselves in the First World War mainly in the area of ideology: with proclamations, manifestations, and articles in the press, acting like protectors of the national culture, they justified and glorified the war. In their "Appeal to the Cultural World" some 93 German intellectuals characterized the war as a battle to defend the German culture, a battle which was forced upon the people.

> It is not true that the fight against our so-called militarism is not also a fight against our culture, as our enemies hypocritically pretend. Without German militarism German culture would have long since been eliminated from the planet. The German militarism has arisen from the German culture to protect that culture in a land which has been ravaged by invasion like no other. The German army and the German people are one.... Believe us! Believe that we will fight this battle to the end as a people with a culture in which the legacy of a Goethe, a Beethoven, and a Kant is just as holy as hearth and native soil (Ref. 4, p. 48ff).

Thus spoke the learned professors of Germany on October 4, 1914—as if that imperialistic carnage had something to do with the legacy of Beethoven! And a philologist of ancient languages in Berlin felt constrained to praise the war as "something grand":

> "Certainly it is something very beautiful and grand for us to have the experience.... Indeed the war is something grand as well, because it takes measure of the soul; it reveals what is in every heart by stripping away all husks of convention..." (Ref. 4, p. 58). A historian was even concerned about "the destiny of the German socialist ideology.... The victory of the Germans also means, from the viewpoint of intellectual history, that German workers will continue to lead. Even the most stubborn doctrinaire needs no proof today that the economic future of the German working class depends on the victory of the German Reich..." (Ref. 4, p. 109).

Scientists on the other side did the same thing—they justified and glorified the war and denigrated the enemy. There was no more talk of the universality and international character of science. There was hardly a voice to protest with

Einstein that "the religious madness of earlier centuries has been replaced by nationalistic madness," nor did many rally behind the socialist Kurt Eisner when he called German science a whore because it surrendered with all its academic dignity and eloquence to the interests of capitalists and government.

Hence World War I, as far as the role of science was concerned, became not just the "war of chemistry" but more broadly the "war of intellects" (Ref. 80, p. 84), in which scholars—regardless of their specialty—made available their high level of prestige for purposes of war propaganda. In the years and decades following the First World War, stabilization of the system as well as justification and interpretation of the values, norms, and institutions of the political system remained one of the principal responsibilities of academic people, who in Germany were mostly funded by the state. The climax in social prestige and self-appreciation for university professors was reached in Kaiser Wilhelm's Germany. With the "decline of the German mandarins,"[162] however, and with the emergence of more effective propaganda techniques, particularly the radio, under National Socialism the academic community lost that exclusive role in matters of war propaganda which they enjoyed during the First World War, and on the other side, the Allies did not need to wage an intellectual war to mobilize opinion against a fascist aggressor.

As the organization and conduct of war became more and more dependent on technology, the roles of specific branches of science became more and more important (Figs. 44–46). In addition to atomic bombs, major collaborative projects between science and the military during World War II produced jet engines, rockets, radar, and computers. To the extent that physics had been given a base in industry, this branch of science assumed a role that was critical for military technology. The Second World War became a war of the physicists. Two prime examples of this have already been described at great length in the literature, namely the American Manhattan Project, which ended with the construction and dropping of the first atomic bomb,[103,64,177] and the German "Uranium Association" ("Uranverein"), which "only" produced a nuclear reactor that was close to functional.[63,96] The spectacular and unique nature of these projects, however, made the interaction of science and the military appear to be something unusual. Depending on the viewpoint of the individual authors, the physicists working on the atomic bomb project were depicted all too lightly as tragic heroes or diabolical geniuses. In order to understand the real role of physicists in World War II and the effects of that activity on the postwar period, it would be more instructive to look at a less sensational scientific military project like the development of radar.

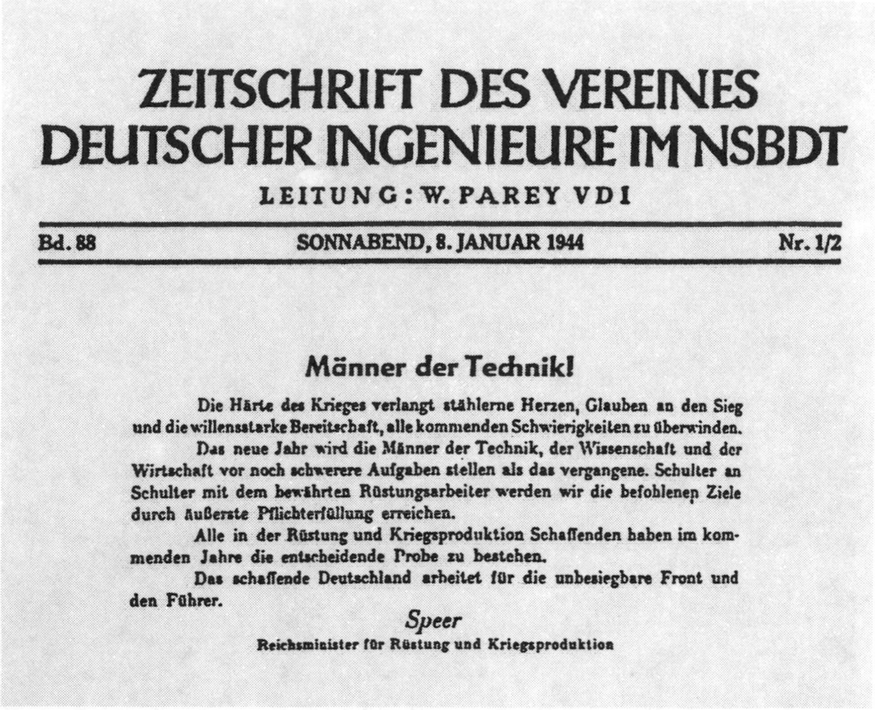

ZEITSCHRIFT DES VEREINES DEUTSCHER INGENIEURE IM NSBDT

LEITUNG: W. PAREY VDI

Bd. 88 SONNABEND, 8. JANUAR 1944 Nr. 1/2

Männer der Technik!

Die Härte des Krieges verlangt stählerne Herzen, Glauben an den Sieg und die willensstarke Bereitschaft, alle kommenden Schwierigkeiten zu überwinden.

Das neue Jahr wird die Männer der Technik, der Wissenschaft und der Wirtschaft vor noch schwerere Aufgaben stellen als das vergangene. Schulter an Schulter mit dem bewährten Rüstungsarbeiter werden wir die befohlenen Ziele durch äußerste Pflichterfüllung erreichen.

Alle in der Rüstung und Kriegsproduktion Schaffenden haben im kommenden Jahre die entscheidende Probe zu bestehen.

Das schaffende Deutschland arbeitet für die unbesiegbare Front und den Führer.

Speer

Reichsminister für Rüstung und Kriegsproduktion

FIG. 44. Nazi propaganda directed to German scientific and technical professionals. [Translation of the page reproduced. Zeitschrift des Vereines deutscher Ingenieure im NSBDT (Journal of the Union of German Engineers in the National Socialist League of German Technical Professionals. Director: W. Parey, VDI. Vol. 88, Saturday, January 8, 1944, No. 1/2. Men of Technology! The rigor of war demands steel hearts, belief in victory, and a clear commitment to overcome all impending difficulties. The new year will make heavier demands than the last on our men of technology, science, and the economy. Shoulder to shoulder with the trusted armament workers we shall achieve the goals set for us by fulfilling our responsibilities to the last detail. All those employed in armament and war production must withstand the critical test of the coming year. Germany's productive resources work for the invincible war front and the Fuhrer. Speer. Reischsminister for Armament and War Production.]

"*Ra*dio *d*etection *a*nd *r*anging", abbreviated "radar," is the name given to a method of locating objects (such as aircraft or bomber targets) using radio waves. The United States invested more money in developing this technology than in the atomic bomb project (about $2.5 billion versus $2 billion). Two aspects of the radar project will be examined, both of which were critical factors in the development of physics into a large science that is indispensable for the state and the economy: (1) the organization of war research, which established

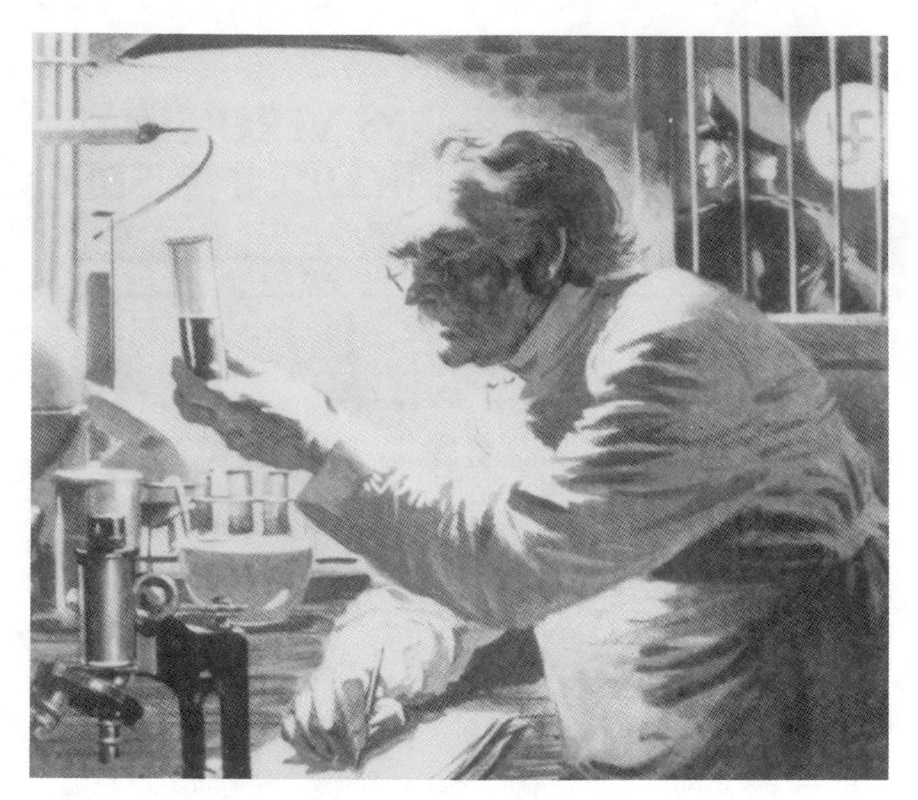

FIG. 45. According to Allied propaganda German scientists were lackeys of the "Huns," carrying on war research in barred and guarded cellars.

contacts among politicians, military people, industry, and science on a scale that was previously unknown, and (2) the semiconductor research which was done under the radar project, which provided the first impetus for the postwar revolution in microelectronics.

THE ORGANIZATION OF WAR RESEARCH IN THE SECOND WORLD WAR

It was clear to all countries involved in World War II that science played a decisive part in that conflict. There were differences, of course, in the scope, timing, and priorities followed in organizing war research. The United States' use of science was the most effective and successful, both in the direct support of the war effort and in its influence on events after the war. Most of the following

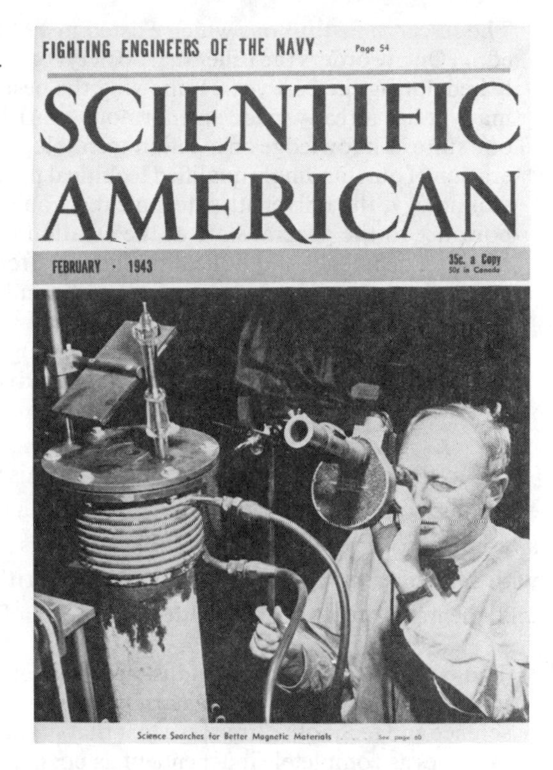

FIGHTING ENGINEERS OF THE NAVY . . . Page 54
SCIENTIFIC AMERICAN
FEBRUARY · 1943
35c. a Copy
Science Searches for Better Magnetic Materials

FIG. 46. The American weapons researcher in 1943 became a front-cover hero of Scientific American.

discussion, therefore, will be concerned with the efforts of the British–American alliance in organizing war research.

SCIENTIFIC PREPARATIONS FOR WAR IN THE NATIONAL SOCIALIST STATE

If German scientists in the National Socialist state were less effective than their American colleagues, the reason lay neither in a lack of determination on the part of those in power to utilize science nor in a lack of willingness to cooperate on the part of most researchers to make their talents available to Hitler's state. Shortly after the Nazis took power they began war preparations with the buildup of an air force, organized and administered by a newly established Ministry of Aviation under the direction of the Minister of Aviation, Colonel General Hermann Göring. In a speech at the opening of the German Academy of Aviation Research on April 16, 1937, Göring left no doubt about the importance of science for the air force:

> The research institutions which existed in 1932 have been greatly expanded... . On my orders the Lilienthal Society for Aviation Research was established during the past year. Following the best scientific tradition, in Germany it has already made an enormous contribution to the enrichment of our store of knowledge on aviation technology. The training of a new generation of outstandingly qualified technical people through the institutes of technology, the universities, technical schools, etc., is of such critical importance to the general state of the aviation technology, that I asked the Minister of Science and Education shortly after we took power to take into account the special requirements of aviation by providing whatever support is needed to solve the problem of recruiting and training new technical staff... . Fundamental and wide-ranging progress in aviation technology can be expected with certainty to flow from advances in physics, chemistry, and other branches of knowledge (Ref. 98, p. 4).

The direct preparations for war by the Nazis were launched in 1936 with an economic policy aimed at autarky. Göring was a key figure in this program also, as commissioner of the Four-Year Plan. Things that he said in a speech in 1937 with reference to the air force also assigned critical responsibilities to physics and chemistry in other areas under this policy of autarky:

> Today the German aircraft industry is still dependent in many areas on the importation of foreign raw materials. As in all major areas of our economy science and technology must see to it that in the aviation field too our Reich becomes as completely independent as possible from foreign imports. This is an area that will continue to demand attention, not just for the next four years, but really forever in the case of German aviation (Ref. 98, p. 4).

Göring was able to recruit famous scientists for his aviation research institutes. Peter Debye (physicist and Nobel Prize winner), Ludwig Prandtl (physicist, specialist in applied mechanics and flow research and director of the Göttingen Aerodynamic Testing Institute), and Jonathan Zenneck (physicist and electrical engineer, president of the Aviation Radio Research Institute), for instance, all belonged to the Committee of the German Academy of Aviation Research. Prandtl also sat alongside Carl Bosch, Nobel Prize winner and president of I. G. Farben, in the chair of the Lilienthal Society for Aviation Research.

In addition to the newly established research institutions in the field of aviation, a series of other organizations was created. The Office for German Raw Materials was intended to conduct mainly chemical research within the policy of the Four-Year Plan and war preparations. In 1937, an Imperial Research Council was set up in which renowned university-level professors were expected to make their various scientific disciplines available to serve the aims of the National Socialist state. There were 13 of these "subject discipline directors," ranging from Professor Esau for physics and machine building to Professor Sauerbruch for medicine. The armed forces also supported their own research facilities. New organizations were still being created in wartime. Thus in about 1942 an Air Force Research Directorate was created, as well as a Navy Central Office for Research and Development Operations.

The efficiency of German war research certainly was not at all proportional to the large organizational investment. A full explanation of this requires an analysis of the scientific and political decision-making processes in the National Socialist state. Since war planning was targeted at first to rapid military conquest (*Blitzkrieg*) and not toward long-term defense efforts, up until about 1942 war-related research that did not promise near-term applications was not supported with the same priority as it was on the side of the Allies. After the failure of the *Blitzkrieg* strategy, it was no longer possible to accomplish a basic readjustment of scientific war research because of the increasing destruction that was taking place.

There was no lack of suggestions in Germany from the ranks of the scientists for the investigation of new military technology. In many areas beside radar (see p. 168ff) and nuclear fission, leading scientists were alerting the authorities to the importance of their research for military applications. The rocket engineers around Wernher von Braun, who even before the war had convinced the Army Weapons Office of the value of their "wonder weapon," gained top priority after the annihilation bombing of Lübeck in March 1942.[108] In contrast to this, top priority was not granted to the development of the atomic bomb either after the first initiatives on the part of several nuclear physicists (1939) or after more than two years of experimental work by the Uranium Association ("Uranverein") on isotope separation and the construction of a nuclear reactor. A recent historical study of the uranium project under the Nazis[198] shows, in contrast to earlier investigations, that the basic reason for this was not the moral position of the nuclear physicists,[103] their inadequate professional competence,[63] nor the ignorance of the responsible political decision makers.[96] In view of the available economic resources it was hopeless to try to produce the necessary explosive material for a bomb through a major engineering project within the projected duration of the war. The project nevertheless was pursued energetically as a scientific undertaking.

CHURCHILL'S SCIENTIFIC ADVISORS

The Allies began relatively late to organize their science for the war. Stafford Cripps, a member of the British War Cabinet, in 1941 said that it was

> I think our main difficulty with regard to the proper utilization of the scientists in this war has been the failure to realize at a sufficiently early stage that this was going to be a truly scientific war and that the battle would not be won merely by the physical ascendancy of our race, but by the ingenuity of those who have been trained in our schools, technical colleges, and universities (Ref. 165, p. 58).

After the end of the First World War, in Great Britain the Board of Defense Research of the Department of Scientific and Industrial Research made a modest start on organizing government-supervised armaments research. This board was headed by a scientist, Sir Henry Tizard, who sought the role of science advisor to the government in competition with the other *éminence grise* of British science politics, the physicist F. A. Lindemann (see p. 90ff). In 1934 Tizard assigned a committee to do a scientific study of air defense, which for Great Britain began the development of radar. Tizard's rival Lindemann joined up with the conservatives and became Churchill's personal science advisor. When Churchill became Prime Minister, Lindemann forced Tizard out of his position as advisor. In 1945 Lindemann followed his boss into the loyal opposition (in the meanwhile he had been knighted Lord Cherwell), and Tizard returned to the helm of scientific political power for the Labour government.

The impetus for further organization of military research came mainly from the scientists themselves. In 1938 the Royal Society, whose membership included the elite of British science, laid a memorandum before the Ministry of Labour, which called for a service obligation for scientific workers during times of national emergency. This resulted in the creation of a central register of scientific and technical personnel who could be drafted in case of war. Sir Lawrence Bragg, Nobel Prize winner and president of the Royal Society, also suggested that students and faculty of universities should make themselves available for service as auxiliary personnel for the research groups associated with the armed forces. Seven weeks before the outbreak of war Bragg initiated the formation of a scientific advisory staff, which was to acquaint the government with "promising new developments of importance for the war effort." Churchill, who wanted no scientists other than Lord Cherwell to influence decisions at the highest political level, assigned a subordinate role to the advisory staff of the Royal Society.

Beyond the central registry of scientific staff, the DSIR, Lord Cherwell, and the subordinate-level scientific advisors, the organization of British war research was based mainly on research facilities of the armed forces (The Royal Aircraft Establishment, the Aeroplane and Armament Experimental Establishment, the Admiralty Research Laboratory, etc.) and on special ministerial committees, such as the Tizard Committee, which played a special role in the administration of the British–American microwave radar research (see p. 141), and the Maud Committee, which initiated and coordinated the first research efforts of the British atomic bomb project.

AMERICA: "BIG SCIENCE" AS A WEAPON

After the Americans entered the war the focus of Allied war research shifted to the United States. Here too it was the scientists who first pushed for effective government organization of war research. Here too a science advisory system was established, which advised the highest levels of government of scientific developments through direct channels, and conversely provided a means of effectively translating political objectives into research programs. And here as well it was aviation that first felt the benefits of strengthened organization of research. As late as 1933 there was still no organization in the United States responsible for overall coordination of civilian and military research.

In 1934 Robert Millikan, the physicist and Nobel Prize winner who was prominent as an ambitious organizer of science during World War I and in the 1920s, when Caltech was being built up, renewed the effort to improve the overall coordination of civilian and military research. He offered the services of the National Research Council (NRC) to the Army at a conference of general officers. Millikan had exercised a decisive influence on the research policy of the NRC ever since its founding during the First World War. The military, however, did not show any enthusiasm for his proposal. They preferred to work with another organization that was also established during World War I and was directly concerned with military requirements: the National Advisory Committee for Aeronautics (NACA), an advisory panel specializing in aeronautical research, with the status of a government agency and the authority to award contracts to both academic and industrial organizations. When Vannevar Bush, another scientist with organizational ambitions, made another attempt in the spring of 1940 to mobilize civilian research for work on military projects, he used NACA as a model for the most powerful war research organization which ever existed up to that time, the National Defense Research committee (NDRC), which later became reorganized into the Office of Scientific Research and Development (OSRD). Responsibility for practically all research involved in the "War of the Physicists" on the side of the Allies—from the atomic bomb to radar—could ultimately be traced to the OSRD.

Bush's career began in Boston, where in 1916 he finished his doctoral work in electrical engineering at Harvard and MIT. In the 1920s his name became known because of his development of a computing device that found application both in science and in industry. His superiors at MIT valued his passion for work, but some of his colleagues considered him arrogant, autocratic, and ambitious, a person who tended to equate the interests of his own computer development program with those of the enitre Department of Engineering. Bush found a powerful supporter, however, in the president of MIT, the physicist Karl Compton, who in the 1930s had developed, as a member of Roosevelt's New Deal, several proposals for a government science organization. Compton made Bush the head of the Department of Engineering and vice president of the university.

Bush felt that the Second World War would become a challenge to natural science and technology, a view he shared with his friends Karl Compton, James B. Conant (the chemist who was president of Harvard), and Frank B. Jewett (the physicist/manager at Bell Laboratories). In 1939 he was made president of the Carnegie Institution of Washington and chairman of the NACA. The organization of aeronautical research, of course, was not enough to occupy Bush. When he saw that the National Research Council was not successful in mobilizing civilian science for weapons research or in coordinating various military research programs in the armed forces, he was convinced that only an independent agency similar to the NACA, with its own budget, established by the government and responsible directly to the President alone, could get the job done. In June 1940 he presented to the President his plan for the establishment of a new government agency called the National Defense Research Committee (NDRC), and Roosevelt did not take long to approve it: "That's okay," the President said after looking for ten minutes at a piece of paper on which Bush had drawn an organizational diagram with four divisions. He then wrote "OK, FDR" at the bottom, as Bush recalled later in an interview. The funds for the new agency were to be taken from the President's emergency fund.

Bush himself became Chairman of the NDRC. Other members were Compton, Conant, and Frank Jewett (the "czar" of industrial research, who in the meanwhile had become president of the National Academy of Science and was promoted by AT&T to a vice presidency), together with Richard C. Tolman, representatives of the Army and the Navy, and a patent expert. Bush himself retained responsibility for dealing with the armed forces, the Congress, and the President. Each civilian member of his panel was assigned responsibility for one of five general sectors of the war research effort: armor and artillery, chemistry and explosives, instrumentation and control devices, transportation and communication, and patents and inventions. Each of them was to be free to establish as many organizational subdivisions as he considered necessary to satisfy the military requirements placed on his division. With this sort of organizational structure the NDRC members had flexibility and could make decisions quickly. In order to avoid rivalries with military research facilities, according to Bush's concept the NDRC was not to replace but to complement existing organizations of the armed forces. Thus Bush also left aeronautical research with NACA. He satisfied the members of the National Academy of Science by according them an advisory role.

The new agency possessed neither its own research facilities nor permanently assigned research staff. Instead it gave money to universities and industrial firms for militarily important research projects. In addition it organized a "National Survey of Scientific and Technical Specialists." "The researchers and engineers of America registered by the thousands, including Albert Einstein, who recalled for the survey officials his own time spent as a patent official: 'I was always interested in practical technical problems.' " (Ref. 111, p. 298).

At the request of the NDRC Karl Compton made the rounds of the military agencies and compiled a list of projects important to the war effort. By December 1940 the NDRC had already let 126 contracts to 32 academic institutions and 19 industrial firms. The following spring a branch of the NDRC was set up in London to coordinate military research among the Allies as well.

Since the Treasury Department required a general report on the organization of defense science one year after the establishment of the NDRC, Bush utilized the opportunity to broaden his authority. On the basis of his report a new agency was established with Bush as director, the Office of Scientific Research and Development (OSRD), which carried on the work of the NDRC but on a larger scale. Up to this point Bush's organization had been concerned only with research; now development was added as well. The agency would no longer be funded from the President's Emergency Fund but from its own budget appropriated directly by the Congress. Bush divided the OSRD into 19 divisions. To the extent possible each subordinate organizational unit would be familiar only with its own area of responsibility. In this way Bush hoped to satisfy the military, which insisted on strict secrecy. Nevertheless, tension rather often arose between the civilian OSRD and the military agencies, but Bush was not worried. He had his own budget, broad authority, and direct access to the President.

> " 'I knew that you couldn't get anything done in that damn town [Washington] unless you organized under the wing of the president,' said the Czar of Research [as the *New York Times* called him] to an interviewer" (Ref. 111, p. 301).

The concentration of power in the OSRD was reflected in the very uneven distribution of projects among its contractors. The largest share of industrial contracts by far went to the Western Electric Company, the manufacturing arm of AT&T. Then followed in order Du Pont, RCA, Eastman Kodak, General Electric, and other industrial concerns whose success even before the war was based on scientific research. Among the universities MIT received the largest number of OSRD assignments, followed by CalTech and Harvard. These institutions were linked closely with the czars of the NDRC and OSRD, Bush (Harvard and MIT), Compton (MIT), Conant (Harvard), Jewett (AT&T), and Tolman (CalTech), all of whom had spent part of their careers there as presidents or vice presidents. The heavy concentration of research money among a relatively small number of universities (90% of the academic research funds went to only eight institutions) and the privatization of the profits realized by industry based on government funding provided by the OSRD (contractors were granted patent ownership for more than 90% of the inventions resulting from the government-sponsored work) made war research in the United States a big business for a small academic and industrial elite, who knew how to make the most of their position of power in the postwar years.

RADAR

Some half-dozen of the 19 divisions of the OSRD were concerned with electronics. A major portion of these efforts were devoted to research and development work on radar. The atom bomb merely ended the war; it was won, however, with radar—this was the way a leading American physicist (Lee A. Du Bridge) summed up the results of his own work and that of his colleagues on this project in an interview.

A NEW VACUUM TUBE FOR MICROWAVES

The first technical breakthroughs on the road to radar were made in England. Although the principle of locating objects by the reflection of directed beams of electromagnetic waves was understood in the 1920s and was tested now and then, successful military applications did not emerge, particularly because of the high sending and receiving power levels required. The concentration of electromagnetic waves into directed beams and hence their effectiveness for radiolocation purposes was easier at shorter wavelengths (if possible, in the range of a few centimeters, that is, in the microwave region), but the capability of available sending and receiving tubes to amplify electromagnetic waves in this frequency range was poor. (A wavelength of 3 cm, for example, corresponds to a frequency of 10^{10} cycles per second or 10 GHz; the transit time of an electron between cathode and anode in an amplifier tube determines the upper limit for the frequency range in which the electromagnetic waves can be converted to electrical currents. With a spacing of 1 cm and a voltage drop of 1000 V the transit time between anode and cathode is more than 5×10^{-10} s. This corresponds to a frequency of less than 0.2 GHz or a wavelength of more than 1.5 m.)

In early 1940 British physicists made a discovery that solved the problem of transmitting tubes for microwavcs. By means of a novel arrangement of anode and cathode in a supplementary magnetic field, they were able to cause heavy electron streams to execute complex movements, the time and space characteristics of which would be controlled by appropriate coupling of the electrodes and the magnetic field. The new tube was called a "magnetron," the essential components of which were a hollow metal tube, the whole of which functioned as an anode, and a cylindrical cathode within it. Conventional tubes, on the other hand, were made of thin filaments of wire in a thin glass envelope, which could not withstand heavy currents of electrons or significant mechanical loading.

A secret committee led by Henry Tizard took this invention to Washington at the end of September 1940 to confer with Bush's NDRC radar experts. The head of the American radar division, which called itself the "Microwave Committee," was Alfred L. Loomis, cousin of the Secretary of War, Henry L. Stimson, and a retired banker. Loomis had specialized in microwave research in his private research laboratory, when the research laboratories of the armed forces concentrated on longer wavelengths, which seemed more readily adaptable for radar purposes using conventional transmitting and receiving tubes. In the summer of 1940 the Microwave Committee was forced to acknowledge that breakthroughs could not be expected in the centimeter-wave range using equipment available at that time. When Loomis learned about the British development of the magnetron from the Tizard committee, he immediately informed his cousin, the Secretary of War, that with this device the U.S. radar program could be pushed forward a good two years.

"RADLAB"—CENTER OF ALLIED RADAR RESEARCH

The Microwave Commitee thereupon decided to established a major radar research laboratory. Bush's NDRC gave the assignment to MIT in Boston, which, in addition to its good relations with the NDRC leadership, enjoyed a good location for conducting radar tests by reason of its proximity to the Atlantic and to the Boston airport. The project was called the Radiation Laboratory (RadLab for short), the same name that had been given to the nuclear physics laboratory under Ernest Lawrence at Berkeley. This was done largely to give the impression to the uninitiated that the research at MIT concerned matters as far removed from military applications as the esoteric field of nuclear physics was perceived to be in 1940. Loomis wanted Lawrence to take over direction of MIT's RadLab, but he declined. He did, however, make himself available to recruit physicists for the project. Lawrence enjoyed the confidence of many prominent physicists and was one of the few who could gather the proper physicists for the project under circumstances in which the recruits could not be entrusted with detailed information about the work until they had actually been hired. "No other method could have provided us with such a select body of physicists. The leadership of a nuclear physicist was needed to recruit them," commented a member of the NDRC about this extremely effective recruiting effort. Lawrence first recruited as Director of RadLab his friend and protegé Lee Du Bridge, who had earned a reputation as an outstanding experimental physicist and administrator while a fellow at CalTech and later as a professor at Rochester University. When a conference on applied nuclear physics was scheduled at MIT for late October 1940, Du Bridge and Lawrence saw to it that those invited to participate included a number of physicists whom they wanted for the radar project. From this group they recruited several future RadLab division chiefs. Those hired often brought along other physicists from their previous working environment. By December 1940 some 50 physicists were already working for the RadLab, including some pioneers in basic research like Edward Condon, Philip M. Morse, Isidor I. Rabi, and John Slater. The organization of the RadLab was—on paper—strictly hierarchical, with committees to set research priorities, divisions, and branches. Many of the physicists on the staff, however, had run their own laboratories in the past and were not accustomed to strict hierarchies based on organizational charts. Hence they developed a flexible system under which priorities were circumvented and researchers switched groups depending on the status of problems to be solved. Since many of the unauthorized research projects ultimately turned out to be useful, the RadLab management tolerated this productive push for independence by the physicists.

The staff grew from about 50 academic scientists in December 1940 to nearly 1200 researchers and 3000 engineers, technicians, and other workers toward the end of 1944. Formally speaking, the RadLab contract was just one of a total of 137 research contracts let by the Radar Division—Division 14—of Bush's OSRD to 18 academic and 39 industrial research organizations. The RadLab, however, played the leading role in the overall American radar project because it was the largest research and development facility engaged in the work. Contacts were established here with various contractors of OSRD Division 14. RadLab signed subcontracts for research with other institutions and these were paid with government funds as contractors of OSRD. Thus MIT's RadLab became the means by which many links were established between academic researchers, industry, the military, and government, links that afforded physicists access to those in power. This alliance was also reflected in the content of the physical research. The science that became the basis for microelectronics, solid-state physics, was given critical impetus by the radar project. This can be seen clearly in the example of a research contract given to Purdue University by OSRD Division 14 in March 1942 with the title "OEMSR 362: Crystal Detectors for UHF."

A RESEARCH CONTRACT FOR PURDUE: CRYSTAL DETECTORS

In the early days of radio, crystal detectors were used as radiowave receivers, but they were replaced by electron tubes as the capability of the tubes was improved. The crystal detectors consisted of a fine metal wire (usually tungsten), which pressed against a polycrystalline semiconductor (often lead sulfide or galena). Because of their smaller dimensions, crystal rectifiers could also be used for wavelengths shorter than those used in radio. Hence they appeared to be suitable for use as radar wave detectors—if they could be made to function reliably. The quality of reception varied depending on the contact between the metal wire and the semiconductor substrate. The search for "sweet spots," where the tip of the wire against the crystal provided optimum reception, was more a matter of skill and chance than of characteristics subject to scientific calculation.

A slight vibration was enough to destroy the effect. This kind of sensitive detector was called a "cat's whisker." In the spring of 1942, when crystal rectifiers were again the subject of research by physicists at Purdue, the crystal detector was still essentially the same unreliable and sensitive "weak link" in a receiver as in the days of the crystal radio (Figs. 47 and 48).

FIG. 47. In using a crystal radio set the operator had to look for the "loudest spot" for radio reception with the tip of a wire.

Purdue University at the time was not as well known as today, and the fact that it was granted an OSRD contract can be attributed to the initiative of the physicist Karl Lark-Horovitz, who in an ambitious and sometimes dictatorial manner had made the physics department of Purdue during the 1930s into a modern research operation. Lark-Horovitz's interest in crystal detectors dated back to the First World War, when he was assigned to operate the crystal radio of his division as a member of the Austrian Signal Corps. After graduating in physics and spending a few years in research at the University of Vienna, he

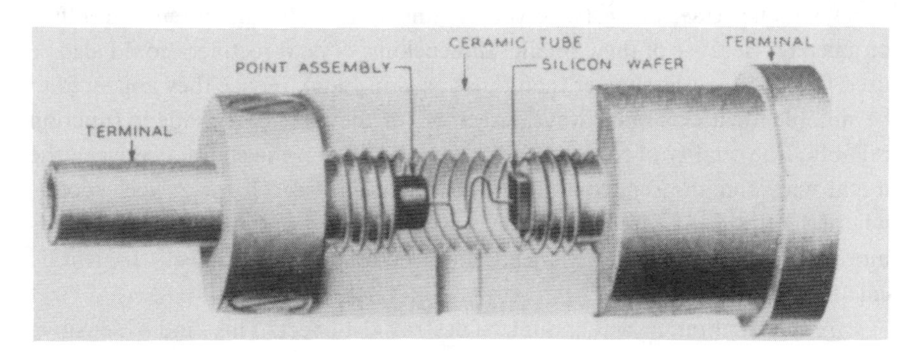

FIG. 48. The first radar detectors for centimeter waves, like the Bell Laboratories silicon rectifier diode shown, were still quite similar to the receiving portion of a crystal radio set. In both cases the rectification needed to detect alternating electromagnetic fields was effected at the point of contact between the tip of a metal wire and a semiconductor surface that allowed electrical current to flow in one direction only.

went to America for a year in 1925 with a Rockefeller fellowship. Purdue University at that time was trying to keep up with the stormy development of physics by modernizing and expanding its physics department, and it hired the Austrian, who was known as an energetic experimental physicist and teacher. His reputation was well deserved. Lark-Horovitz instituted within a short time modern solid-state investigative methods such as x-ray and electron structure analysis; he outfitted a laboratory of spectroscopy and had a cyclotron built for the production of isotopes and the conducting of nuclear fission experiments. As head of the department he was researcher, teacher, and administrator all in one. His zeal for work was inexhaustible, and he expected the same commitment from his students.

> He wouldn't leave this place [Purdue] for any length of time, day and night for that matter, recalled one of his first students, I mean you'd better be on the job at least by 10 o'clock in the morning, because that is when Larkie—as we called him behind his back—came in... . And you'd better be there on top any time until at least about midnight, when he would make his kind of final informal rounds... . And that was the scheme from when I first came here, up until 1958... .[75] Lark-Horovitz died on April 14, 1958, of a heart attack while working in his office.

The crystal detector research contract came about when a Purdue physicist was invited to the RadLab in late 1940 to act as a "general problem solver" among different research groups in tackling new problems and to look for broad solution possibilities that transcended the individual groups. At the end of 1940 Lark-Horovitz asked "his" Purdue man at the RadLab whether he could not get him an OSRD contract in connection with the radar project. He thereupon compiled a list of open questions that could be investigated outside the RadLab. Two topics on this list concerned the properties of crystal detectors, and thus were problems that could be tackled with the experimental solid-state research methods of the Purdue physicists. Lark-Horovitz wrote the following in this application to the RadLab for his own OSRD project.

> "I have had experinece with the making and testing of crystal detectors at broadcasting frequencies." At the same time he laid out a research plan in which the individual steps of the investigations to be undertaken were outlined:
> (1) Construct a 10-cm emitter, intensity controlled by a standard detector; (2) investigate various detector combinations and detector mounts for sensitivity, shock proofness, and capacity; (3) investigate sensitive and insensitive crystal faces by electron diffraction and electron optical investigation to see if any structural effect can be established; (4) investigate the contact microscopically to distinguish between sensitive spots on the crys-

> tal due to impurities and geometry of the contact; (5) produce crystal powder cartridges to allow the use of pre-fab mounts; (6) artificially produce insulating layers to try to make a face-to-face rectifier instead of a point-face rectifier.[75]

The RadLab immediately approved Lark-Horovitz's project, and the Purdue physicists were able to begin work on it in March 1942.

The group around Lark-Horovitz consisted of a nucleus of eight physicists. The oldest among them had obtained his doctorate in 1934, while the second obtained his degree in 1941. The younger members of the group had just finished their comprehensive examinations for the physics degree. The rest of the staff consisted of senior-level students. Overall the group had but little experience in the field of semiconductors. Since initially the crystal detector problem consisted more in the production of an optimum contact between the substrate and metal (according to the plan) than in a study of the semiconductive properties of the substrate itself, this lack of experience did not seem to be particularly important. Perhaps a better understanding of the problem would not, however, have made any difference in the composition and competence of the group. The state of semiconductor science, as set forth, for example, in Frederick Seitz's 1940 book entitled *Modern Theory of Solids*, would not have been sufficient in any way for building proper semiconductor detectors.

GERMANIUM, A NEW DETECTOR MATERIAL

Lark-Horovitz's group rapidly concluded that the initial formulation of the problem was intractable. Instead of studying the contact properties using the conventional detector material lead sulfide, as stated in the research plan, the Purdue physicists concentrated on a new substrate called germanium. A critical factor in that decision was perhaps a visit by Lark-Horovitz to the Sperry Gyroscope Company, where tests with this substance had just begun. Work with silicon, which also appeared to be a suitable detector material, was already under way with other RadLab contractors, including Bell and du Pont, where Seitz was working on the production of pure silicon. "Larkie's" group therefore focused their attention on germanium, which was not as well known, and a quick study of the literature showed that pure germanium should possess ideal rectification properties. Why germanium behaved in this way was unknown. It was not even known that in germanium one was dealing with a semiconductor.

The lack of theoretical knowledge, however, was no obstacle as far as the direct war needs were concerned. The object was the production of radar detectors, not the advancement of basic solid-state physics. Hence the physicists at Purdue did not notice that they were also close to an epoch-making discovery in their resistance measurements in the vicinity of the germanium-metal contact. The resistance was much smaller at this site than in the rest of the germanium because of the injection of charge carriers from the metal wire into the semiconductor. But the "discoverers" of the effect had no inkling that here they were looking at the principle of the transistor.

With the choice of germanium the Purdue project assumed an orientation different from that originally envisioned. The first goal was to produce the pure substance, since it was not readily obtainable. This required close collaboration with the chemical industry, which prepared germanium as a by-product in the processing of lead as germanium tetrachloride ($GeCl_4$) and offered it for sale. Randall M. Whaley, who worked on this problem as a physicist from the Purdue group with chemists from the Eagle-Pitcher lead factory in Joplin, Missouri, in the winter of 1942/43 succeeded in producing very pure germanium crystals as well as germanium with controlled impurities, so that the material could be used to build detectors with unusually high inverse voltage (i.e., the voltage up to which currents in the direction opposite to the conducting direction are blocked). It was precisely this property that was of great importance in view of the often unpredictable high voltages that sometimes occurred in the radar devices of that time. Most of the experiments were conducted according to a strategy of systematic testing and not a logic determined by theory. Whaley, in his search for diodes with a high blocking voltage characteristic, tried almost all materials available to him as admixtures in the germanium. Seymour Benzer, who had just completed his preliminary examinations for his doctorate in physics in 1942 and who was entrusted with routine conductivity meaurements in the inverse direction, discovered by chance during "overloading" of his diodes in the inverse direction that the contacts between the metal wire and the germanium substrate were strengthened in a useful way and the diode continued to function as a rectifier.

On May 1, 1943, Benzer reported at a conference of RadLab researchers concerned with detector problems ("a crystal meeting") on the "burn-through" experiments and the high inverse voltages diodes made of germanium and germanium alloys. His reports were initially greeted with skepticism, since control experiments did not lead to the same results. It was not until early 1944 that sufficient test results were available to begin mass production of germanium diodes. Western Electric undertook production, while Purdue conducted the necessary tests. With the military application of germanium in radar equipment, this long-neglected element, which received its name at the end of the

19th century by a patriotic chemist from Saxony, became a weapon whose effect was drastically expressed in the enhanced, weather-independent accuracy of bombing attacks over this same "Germany." Thus began a germanium boom, which continued after the war was over. No other semiconducting substance could be produced with such purity; measurement results for electrical conduction properties were available for a wide variety of admixtures of foreign atoms, and these measurements could serve as the basis for the start of solid-state electronics.

After the war Purdue was no longer some insignificant provincial university in Indiana. The military had come to understand and value the relevance of physical information for its purposes and became, through the Army Signal Corps, a new financial support organization for Purdue's semiconductor research, where after the war, when the pressure for directly useful results was removed, basic research was more strongly emphasized than applications. Lark-Horovitz became a "pioneer of solid-state physics,"[101] and this was the basic science of microelectronics, which would signal the start of a new industrial revolution.

MILITARY RESEARCH AT BELL LABS

Approximately half of the total expenditures for war research at Bell Labs was for radar work. Although the newly started research program in the area of solid-state physics at Bell Laboratories was cut back sharply at the beginning of World War II, after the war Bell became the largest research facility in this field. This development was started by a general expansion of research as a result of military arms contracts. Table VII shows the amount of effort that Bell labs expended on military projects as an industrial prime contractor of the OSRD. The contracts from Bush's organization, which numbered more than 1500, covered, in addition to radar studies, research on wireless communications for military applications (e.g., receivers that could be operated under a gas mask), gun fire control, underwater sound, and rockets (Figs. 49–52). In 1939 the total amount of government contracts was approximately $100 000 or 1% of the total Bell Labs research expenditures. In 1944 this sum had risen to $56 million and constituted 81% of the research expenditures. The technical staff increased tenfold between 1940 and 1943. The work week increased on the average to 66 hours, and a 90-hour work week was not unusual. Vacation time was limited to two days per year.[37]

Approximately half of the military research expenditures at Bell Labs were for radar work. As early as 1937 engineers from Bell Labs had been entrusted

TABLE VII. Assignment of Bell Laboratories technical staff.

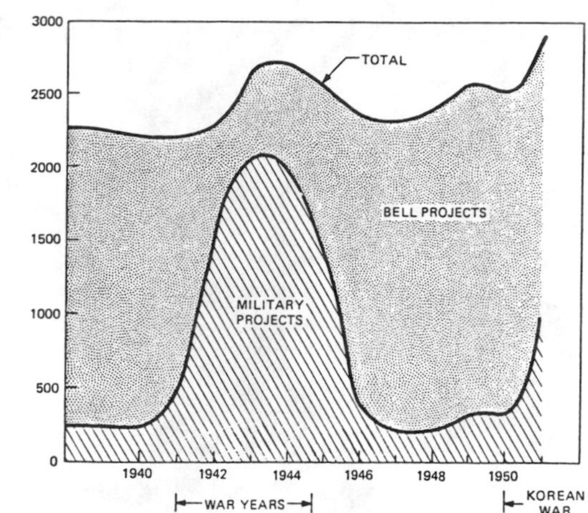

with the radar secret during a visit to the Naval Research Laboratory. A contract with the government was negotiated, and two years later Bell had its first research model of a radar device, which operated in the 60-cm wavelength region. Devices of this sort were used by the Navy in the South Pacific during the early years of the war. In April 1942, Western Electric, the production arm of Bell Laboratories, delivered the first microwave radar for use in submarines. In addition, portable radars were developed for the army to locate mines; radar equipment was produced for the Air Force for recognizing bomber targets; and ground-station radar facilities were developed for warning of the approach of aircraft and for tracking the flight path of projectiles.

The radar project not only brought government research contracts to Bell Laboratories on a scale previously unknown, but it also assured Bell Labs of the continuing esteem of the military (Fig. 47). In a congratulatory telegram dated January 12, 1944, the chief of staff of the Air Force, General H. H. Arnold, sent the following message to the president of Bell Laboratories:

> Directly as a result of the outstanding contribution made by your organization in the development of special electronic equipment and in the making of preproduction models thereof, it has been possible for the Army Air Forces to take the offensive with telling effects against Japanese shipping in the South and Southwest Pacific areas at a much earlier date than would otherwise have been possible and under conditions which normally would have made such operations impossible.
>
> It is my great pleasure on behalf of the Army Air Forces to express our appreciation for this contribution to congratulate you and your people on their achievement.

FIG. 49. "We're backing them up." With these proud words Bell's telephone man hurries to meet the needs of land, sea, and air forces in an advertisement from the year 1942.

FIG. 50. In this advertisement of 1943, the Bell System asks the American public for understanding in the face of difficulties experienced by the civilian telephone system as a result of war-caused shortages and the call-up of many Bell employees to the armed services.

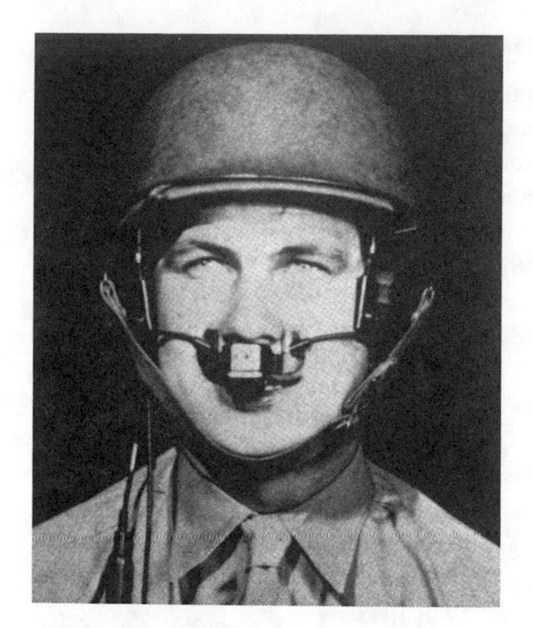

FIG. 51. The Bell Laboratories lip microphone, which was used by land and sea forces because of its low noise level.

FIG. 52. A bell radar set for directing the on-board cannons of fighter aircraft. Enemy bombers could be located with it at a distance of 16 km.

THE ORGANIZATION OF RADAR RESEARCH IN GERMANY

German military research in the field of microwave radar was far behind that of the Allies, even though the existence of this technology was made known quite early. In 1937 Professor Esau pointed out the importance of "electrical waves in the centimeter region" in a lecture to the German Academy of Aeronautical Research. "The technique is concerned not only with the identification of stationary objects through the reflection of centimeter waves but also moving objects, such as aircraft, for example" (Ref. 35, p. 14). These words of Esau also pointed to the type of application for microwave radars. Even though there was no lack of industry experience in high-frequency matters (Telefunken, Siemens) nor of competent physicists and engineers, radar research in the centimeter wave range was neglected in Germany, while the English and Americans devoted great effort to precisely this area. German radar research was uncoordinated and split among research laboratories (mainly Telefunken), individual institutes at universities, and naval and air force research centers. In a statement of accounts of the "High Frequency Research Authority" appointed by Göring, Dr. Hans Plendl had the following to say in 1943:

> German research capacity in this field is about 10 times less than that of the enemy and is still divided up among about 100 different laboratories, which are mostly small to very small. The research capacity of foreign countries (Netherlands and France), which has basically been available, has for practical purposes remained unused. As a result of this splintering of effort, most of the institutes and laboratories have not been able to work on the really large problems of electronics in warfare or have not been able to do the job adequately... . The research assignments from the various organizations requiring such work were given out without adequate coordination. There was hardly an exchange of information among the various institutes, so that frequently there was considerable duplication of work. These deficiencies and particularly the lack of practical experience relative to problems important to the war effort resulted in a situation where the electronic industry was forced to undertake development work in particular areas without adequate preliminary research (Ref. 161, p. 193).

The superiority of Allied radar research began to have very noticeable effects on the course of the war of 1943. The rapid increase in submarine sinkings by British bombers was the first compelling evidence of this. Using a 9-cm radar set, which had a very high transmitting power by virtue of the new magnetron tube, the bomber pilots could clearly recognize submarines on their display screens at night through clouds and fog. About 200 submarines were sunk in this fashion (Ref. 16, p. 42). One of these radar sets fell into the hands of the

German military in February 1943 from a bomber that had been shot down over Rotterdam. A "Rotterdam Study Group" was quickly established to reveal the secret of this dented and blood-smeared instrument. In July 1943, under a project labeled "Rü-Funk," 1500 specialists were detailed from their units and assigned to research and development work in the centimeter wave range. At the same time (July 24, 1943) British bombing squadrons mounted the heaviest air attacks up to that time against Hamburg—at night and with devastating precision! None of the numerous books that glorified wartime technology has neglected to cite this example showing the effectiveness of radar. One of these literary efforts, of questionable merit in spite of considerable technical expertise, describes the moment when Telefunken engineers finally succeeded in copying the "Rotterdam" device:

> Göring now ordered that the laboratories be moved for future work into one of the strongest antiaircraft towers in Germany, that in the Humboldt Woods in Berlin... . Building the copy exceeded all demands which had been placed on radar technology up to that time in the war. The 9-cm tube which was put into service was the magic crown jewel of all vacuum tubes, the "magnetron,"...
>
> Work in the antiaircraft tower did not let up. When the tubes were finished and installed, wires in the experimental apparatus began to char, short circuits occurred, and the detector could not withstand the heavy overload. Additional cutoff circuits, interruptors, and other circuits had to be installed. One August day in 1943 the equipment was ready.
>
> It was taken to the top platform of the antiaircraft tower, switched on, and the antenna began to rotate. The head physicist adjusted the knobs and he turned pale as he stared at the screen: the city of Berlin appeared on the screen with its principal streets, residential buildings, parks, and lakes. The screen showed more than was visible to the human eye from the tower. Because of their "grazing incidence" the centimeter waves penetrated everywhere, into the streets, into the courtyards within a radius of 70 km, and they faithfully reflected all outlines with a greenish phosphorescence. The Wannsee was clearly recognizable in all its details in the left of the screen; the Müggelsee lay to the right... . The engineers on the antiaircraft tower knew what this meant. Göring and navy headquarters were notified immediately. Göring sent an adjutant; naval officers appeared later.
>
> Göring's adjutant carried a grim message back to headquarters. The Allied air forces were able to see German cities, villages, industrial sites, and the attack targets among them, at night and under foggy conditions as if they were being shown on film. It was as if scales fell from the eyes of the men on the antiaircraft tower: the attack on Hamburg had yielded its last secret (Ref. 13, p. 22ff)

The SS was also active in the field of radar research as well as in the development and production of electronic warfare equipment. At the Dachau concentration camp, Oswald Pohl, head of the "SS Economic Administrative Headquarters" and Himmler's senior group leader and general of the *Waffen-SS*, established a high-frequency institute and assigned its "scientific direction" to the former director of the Siemens & Halske Central Laboratory, who had been incarcerated because of statements concerning the Nazi regime. Later on this "institute" was moved to the Gross-Rosen concentration camp, where in early 1944 some 200 educated prisoners, in four barracks built especially for this purpose, were to undertake the production of measuring instruments and other high-frequency equipment (Ref. 117, p. 490). According to a tally dated February 21, 1944, concerning the assignment of concentration camp prisoners to the armament industry, 935 inmates were utilized on electronic projects in the aviation industry, representing a monthly labor investment of 258 742 hours, the majority of which were used in the "manufacture of intelligence devices" for the Siemens & Halske firm in its branch at the Ravensbruck concentration camp (Ref. 114, p. 112ff). For this the collaboration of the SS and industry was in no way unilaterally dictated by the SS. The SS sent concentration camp inmates for labor services to industry only at the request of interested enterprises. In an order issued to all camp commandants, an *SS Obersturmbannführer* from the SS Economic Administration Headquarters specifically emphasized "that requests for prisoner assignment are to be submitted to me for approval *prior* to segregating a detachment" (Ref. 58, p. 146).

Although the use of prisoners in industrial plants of concerns like I. G. Farben and Krupp have long been known around the world as a result of the Nuremburg trials, the involvement of electronic firms like Siemens, AEG, and Telefunken in the slave labor system has only been known publicly for a short time.

> Although none of their directors [of AEG, Telefunken, and Siemens] stood trial in a Nuremburg court, they too joined in a partnership with the SS when they went to Auschwitz, Ravensbruck, Dachau, Mauthausen, and other death camps to obtain workers. This statement was made in a historical review concerned with the "wages of horror" and the "compensation denied to Jewish forced laborers" (Ref. 38, p. 21).

It took the research and proof developed by a Jewish organization (the Claims Conference), conducted over decades on a worldwide scale, to induce Siemens, Telefunken, and AEG to finally pay damages of a few million marks.

CENTIMETER WAVE DETECTORS

In Germany, as in the United States, natural crystals like pyrite and galena, which were familiar from the days of the crystal radio, were utilized at first as semiconductor radar detectors. Toward the end of the war detectors made of synthetic germanium were produced in the Siemens branch plant in Vienna. Telefunken produced detectors made of artificial pyrites with synthetic layers of silicon at relocated production facilities in Thuringia.[174] The Siemens detector was suitable for use only to a small extent prior to the end of the war in the range of wavelengths below 10 cm as used by the Allies. Telefunken, however, was able to use a makeshift detector in submarine warfare. This detector was the so-called "Naxos finger" ("Naxos" was the name of a rather old radar receiver that was equipped with a 9-cm finger-shaped detector).

> In an emergency a drowning man clutches at straws. In spite of its shortcomings the detector device was in such demand that submarine commanders went after them to Berlin from the harbors on the French coast and would not return until they had a receiver in hand.
> This equipment had a range of only 9 km. This distance represented an hour and a half of travel for a hiker. An automobile covered the distance in 9 minutes. An enemy aircraft, however, required only 60 seconds or less to travel this distance. The battle against death had become a matter of seconds. A submarine required 30–40 seconds to dive. In order not to waste even an instant, the radar operator stood next to the commander in the tower. He held the Naxos up in the air for hours at a time and watched it carefully, with his arm outstretched to avoid inductive interference. Reception was not equally good in all directions. Hence the operator turned continuously with the device in his hand. The dial registers! The magic blip appears and shows an enemy machine: "Alarm...dive!" The command was obeyed in seconds. Was it already too late?
> It often happened that antiaircraft crews on deck did not reach the tower in time. Mate Jessen and Cannoneer Kraus are still outside. Those two or the enitre boat, two men or forty, that is the question, and it is decided with lightning speed according to the relentless law of war. The two antiaircraft crew members will never again return to the tower. They are washed away by the next or a later wave... (Ref. 13, p. 24ff).

In an article in the *Telefunken Zeitung* in 1950 entitled "Technical Development and Research at Telefunken During the War" that firm's microwave research was summarized in rather restrained terms as compared with the self-confidence of Bell Laboratories:

> In the later days of the war the frequency region of German radar was also shifted into the centimeter wave range following the example of the English

> and Americans. Entry into this range was made easier by study of captured equipment. After experience had first been gained by copying such equipment, domestic development became very rapid, adapting this technology to the different requirements of the German air force (search and guidance equipment for night fighters). For example, the development of highly sensitive receivers operating on pure alternating current was entirely original... (Ref. 118, p. 19).

A number of engineers and physicists who worked on radar research during the Second World War later played an important part in the development of electronics in the German Federal Republic: Leo Brandt, head of equipment development at Telefunken and chairman of the "Rotterdam Working Group," assumed a leading role in research policy in the 1950s and 1960s. The physicists Kurt Seiler and Heinrich Welker, who were critical players in the radar detector research at Telefunken and Siemens, in the 1950s built the first industrial semiconductor research laboratories. It was the United States however, which provided the decisive thrust for postwar development in electronics. Except for a few special areas (such as, e.g., III–V compounds; see p. 193ff), after World War II the dynamics of industrial research in the electronics sector throughout the world was determined by development activity in the United States.

5

SOLID-STATE ELECTRONICS—A NEW INDUSTRY

THE RADAR PROJECT LED first of all to "knowledge about microwave transmission, waveguides, antenna design, and the development and application of radio tubes" (Ref. 37, p. 131). The work in the research laboratories was not concerned with scientific knowledge, which could be utilized only years later, but with technology that could function as rapidly as possible, even though the laboratory work was done by scientists. This was true even for groups working with new semiconductive materials as radar detectors. The physicists began to study the scientific foundations of the new technology only after the war.

BELL LABORATORIES PLAN AN INVENTION

The numerous possibilities for using semiconductors in the 1950s led to an expansion of the electronics industry, and there is still no end in sight. This development was introduced with the transistor, a semiconductor component that does the work of an electron tube but requires only a fraction of its volume and electrical power. (See the Appendix, p. 212ff.) Its invention in the year 1947 was the result of a research program which Bell Laboratories used to establish the scientific underpinnings for the semiconductor experience gained in the radar project.

"SUBJECT: SOLID STATE PHYSICS"

The search for microwave radar detectors led the Bell Laboratories physicists to the semiconductive material silicon, which even before the war was used to make point-contact diodes ("cat's whiskers"; see p. 143ff). Brattain and

Becker investigated detectors of this sort for radar wavelengths between 1 and 10 cm. Advances could be made quickly with metallurgists, chemists, and specialists from other Bell Labs divisions. The surface of the silicon could be prepared in such a way that it was no longer necessary to search for "sweet spots" for contact with the tip of the metal wire, a search that previously was so tedious. Improved design reduced the effect of external factors such as shock sensitivity and moisture. It was not necessary to understand the physics of what went on in these diodes to make such improvements.

Progress with the use of germanium and silicon as detector materials during the war brought commercial applications of these substances to the attention of research planners, who were already thinking about the postwar period. Bell Laboratories' research director Kelly was giving particular attention to applications in the field of radio for commercial exploitation after the war, once the field of microwaves and radio waves had been opened up to his organization through the radar project. As early as 1943 he had developed a plan for reorganizing Bell Labs research in order to be able to cope with the expected difficult competition situation on a scientific basis. He felt it was urgent to expand the laboratories at Murray Hill in New Jersey, which had become involved in war research as early as 1941. In order to recruit competent scientists, he proposed a further liberalization of the firm's attitude toward fundamental research, along the lines that had previously been followed in the 1930s. The contact which became necessary during the war between the scientific groups and the development and construction divisions was to be maintained, as was the high standard in the selection of staff. The physical research division was reorganized in July 1945 in order to give more emphasis to solid-state physics. Of the nine research groups three were to be concerned exclusively with the foundations of solid-state physics: "physical electronics," "dynamics of electrons," and "solids" in general. The first work plan of the "solid-state group" contained the following statements:

> Subject: solid-state physics. Basic research on conductors, semiconductors, dielectrics, insulators, and piezoelectric and magnetic materials. The research work under this project will develop new fundamental knowledge on the utilization of completely new and improved components and devices for communication systems... . We see great opportunities to produce new and useful properties through the discovery of physical and chemical methods for controlling the arrangement and behavior of the atoms and electrons which make up a solid (Ref. 200, p. 71).

Five subgroups of the solid-state group reflected the firm's areas of interest. They were interdisciplinary in composition and carried on research in the fields of magnetism, contacts and carbon microphones, photoelectricity, crystals for oscillators, and the physical chemistry of solids and semiconductors. The work of the semiconductor group led ultimately to the discovery of the transistor.

THE SEMICONDUCTOR GROUP

Shockley, who returned to Bell Laboratories from his war service at the Pentagon in 1945, took charge of the semiconductor group as a theoretical solid-state physicist. The rest of this group consisted of the experimental physicist Brattain, who had been studying the properties of semiconductive surfaces since 1931; Pearson, also an experimental physicist, who had spent many years studying the interior properties of semiconductors; Hilbert Moore, a specialist in electronic circuits; and Robert Gibney, an expert in physical chemistry. To round out his team Shockley suggested to his superior Mr. Kelly that a position be offered to John Bardeen. Bardeen had earned his doctorate at Princeton under Wigner and had worked with van Vleck at Harvard, where Shockley had met him. Both Wigner and van Vleck were among the leading theoreticians in the field of solid-state physics. The general charter of the newly formed group called for the development of fundamental knowledge on the properties of semiconductors, but everyone had "in the back of his mind the objective of building an amplifier out of semiconductive material."[6]

The semiconductor group at Bell Laboratories began its work with a survey of wartime developments in the field of semiconductors in their own research laboratories and those of others. Thus Shockley and Morgan visited the laboratory of Lark-Horovitz at Purdue University to familiarize themselves with the results of research on germanium. The team decided to cut back the investigation of silicon and germanium. The crystalline structure of these materials was substantially simpler than that of the cuprous oxide and selenium used earlier. In addition, experience was gained in the course of wartime research with the production of these crystals and their doping.

Shockley now resumed his solid-state research at Bell Laboratories from the year 1939 (see p. 118ff) which had been interrupted by his war work. He sketched in his log the construction of a "field-effect amplifier." The movement of current through a semiconductor was to be controlled by a small change in the number of charge carriers. A small metal plate on the surface was to attract or repel the charge carriers depending on polarity. With this method it was intended to make the current flow variable over a wide range and result in amplification of an alternating current in the crystal through a periodically applied voltage. The idea appeared convincing.

Shockley could not patent the idea, however, since Julius Lilienfeld had applied for a patent in 1926 on a very similar process, which, however, was never put to practical use. The experiments, however, in which Brattain collaborated, could not demonstrate the predicted effect.

In order to understand the failure of the experiments, Shockley had to assume that the increase in conductivity actually had reached barely 1% of the theoretically predicted value (based on Schottky's surface layer theory). Shockley had Bardeen check his reasoning. Bardeen conjectured that the induced charges traveled only on the surface and were captured there at fixed sites and thus were no longer available for conduction of electricity. He expanded the models of the semiconductor rectifiers by postulating surface states resulting from irregularities in the crystal structure. Even less than ten such defects per thousand surface atoms were sufficient to shield the interior of the layer completely from the electrical field.

By making this assumption it was possible to explain additional observations that did not agree with the Schottky theory. Contrary to expectations the properties of a metal-semiconductor detector did not depend on the type of metal used. The stationary surface charges blanked out the differences in the different kinds of metals. The lack of success in building Shockley's field-effect amplifier had thus led to a new physical theory and directed the interest of physicists more strongly to the surface of solids. Pearson, Brattain, and Bardeen during the next 20 months pursued an extensive program of investigation of surface states. Relationships to spatial properties such as velocity of holes and electrons in semiconductors were investigated as well as the influence on surface properties and contact voltages. In order to be able to study conditions on the surface it was necessary to have measuring probes which provided exact information on fields, charges, currents, and voltages. The surface was scanned with the tips of very tiny metal wires. The setup was reminiscent of early crystal detectors. In contrast to the undefined conditions involved in the use of "cat's whisker" materials like lead sulfide and selenium, the researchers from the semiconductor group now were experimenting with extremely uniform crystals of silicon to which impurities were deliberately added.

A decisive series of experiments began in November 1947. Brattain had noticed by chance that the measurements were sharply different when water had condensed on the surface. In all probability mobile ions created an electrical field in the liquid and the field overcame the shielding effect of the surface conditions. On November 20 Gibney and Brattain wrote a patent specification indicating that this effect could make possible the construction of a field-effect amplifier. A few days later Bardeen submitted a sketch of an amplifier arrangement. A fine point of metal pressed against a piece of silicon. The pressed semiconductor was to be doped in such a way that the current was carried by positive charge carriers. Negative electrons would carry the current only in a thin surface layer. The effect of surface conditions was removed with an electrolyte—in this case a drop of water. On November 23, 1947, Bardeen wrote in his laboratory notebook that it was possible to amplify the flow of current from the electron-conducting layer into the metal wire tip by applying a voltage between the drop and the block of silicon. But no change occurred in the voltage between the block and the wire tip. This design, of course, was still entirely unsuitable for practical application. The drop of water evaporated quickly and only slowly changing currents were amplified.

THE POINT CONTACT TRANSISTOR

The semiconductor group decided to use germanium in future experiments, since this material was available in higher purity. Also, the drop of water which was occasionally present in the previous experiments was replaced by a suitable electrolytic substance. In the new arrangement the amplification was actually higher, but the frequency behavior remained unsatisfactory. Bardeen later reported:

> "We were not able to make the thing amplify at frequencies higher than 10 Hz. We assumed that the reason for this was the sluggish reaction of the electrolyte" (Ref. 88, p. 65). Again a more casual observation provided further help.
>
> We found that the effect increased when a voltage was applied to the electrolyte for a period of time... . We could also see that a layer formed under the electrolytes. This film must have acted as an insulator. In this case a metal electrode can be applied to the film in order to obtain the field effect even without an electrolyte and to achieve higher frequencies (Ref. 88, p. 65).

A thin layer of gold was vapor deposited in order to help control the number of charge carriers. Since they were concerned in this experiment with positive charge carriers, a negative voltage should have led to an increase in holes and hence to amplification. Again the results baffled the Bell researchers. The expected amplification did indeed appear but only with a polarity that should have driven the positive charge carriers out of the semiconductive layer. Careful examination of the crystal showed that the insulating oxide layer was washed away with the electrolyte. Gold and germanium therefore were connected conductively and the positive charge carriers must have flowed from the positively charged gold layer into the germanium layer. These injected holes, as they later were called, influenced the flow of current from the tip of the tungsten wire into the germanium. The effect of one contact on the other must have thus been carried through the crystal even though the distance between the contact amounted to about 0.1 mm, a long distance measured compared to the free path length of the charge carriers in the crystal. The injected charge carriers played a role similar to that of the grid in a triode. There could no longer be talk of a field effect. Brattain and Bardeen noted that this entirely different principle could be applied to the design of an amplifier. Bardeen later conjectured that without the inadvertant removal of the oxide layer the further work might well have led to the discovery of the desired field-effect amplifier which has come to be used in modern planar technology.

The further experiments in December 1947 were concentrated on the point contact. Bardeen reckoned that a significant amplification would be achieved when the contacts were spaced only some 0.05 mm apart. At that time this was a very difficult problem, which Brattain solved elegantly. He applied gold foil to a three-dimensional prism and cut it on one side of the triangle carefully with a razor blade in order to obtain two closely spaced contacts. This arrangement worked on the first try on December 16, 1947. It amplified signals up to 100 times, and amplification occurred even at frequencies in the radio wave region (Figs. 53 and 54). A step had been made toward Kelly's research objective in

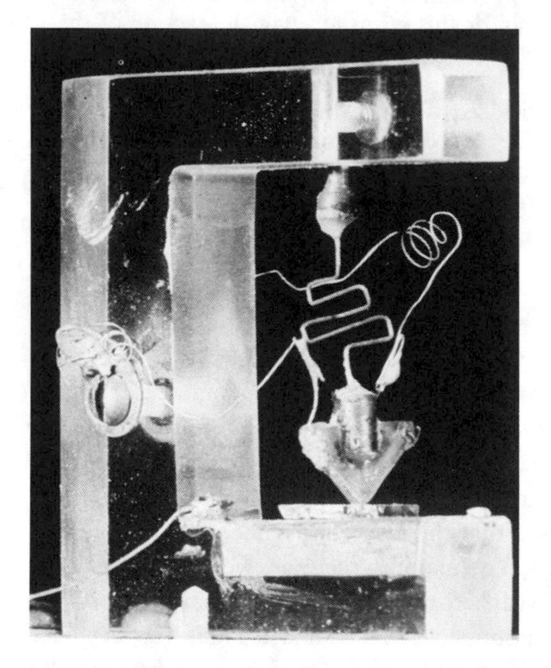

FIG. 53. The first point-contact transistor. Gold foil was applied to the edges of the plastic triangle, and the gold foil was slit on the superimposed point with a thin razor blade to make closely spaced contacts. A coil spring pressed the contacts against the semiconductor surface.

FIG. 54. Diagram of the transistor setup. The collector current I_c could be controlled with voltage V_1.

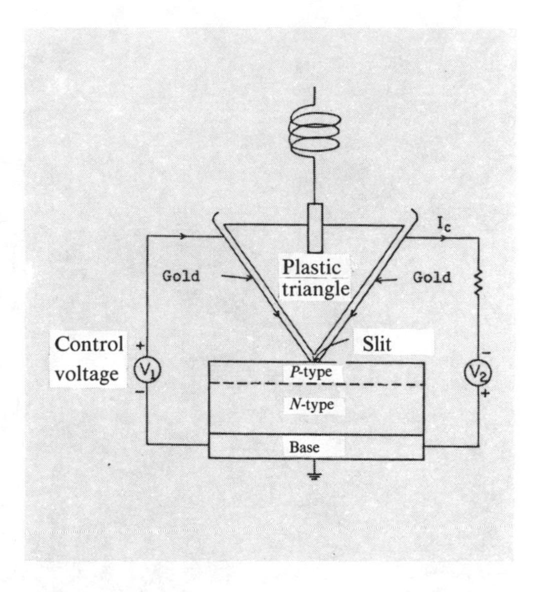

the 1930s, the development of a practical solid-state repeater. A week of intensive work followed to make sure that no errors crept in. On December 23, 1947, the amplifier was shown to management. One day later Brattain wrote in his laboratory notebook:

> The device was installed in a circuit. A voice signal was applied to this circuit and by switching it on and off a clear amplification of the voice volume was heard and seen on the oscilloscope without noticeable deterioration of quality. Measurements at constant frequency showed an amplification performance of about 18 or more.... This demonstration took place on the afternoon of December 23, 1947 (Ref. 88, p. 70).

This day is viewed today as the day of the discovery. Later the official photographs of the three inventors were taken. Bardeen and Brattain watched while Shockley looked at the amplifier through a microscope (Fig. 55). A name had to be found for the new electronic component. It should fit in with the series of other semiconductive components like the thermistor and varistor. John Pierce, another Bell staff member, wanted the name also to note the transfer of charge carriers *through* (trans) the crystal and suggested *trans*istor. "That's it!" answered Brattain. Bell guarded the invention for seven months as a laboratory secret. The staff was not to mention the transistor outside the group. On June 22, 1948, the firm revealed the secret to all the technical staff and then presented the invention to the military. For a whole week the military was the only group

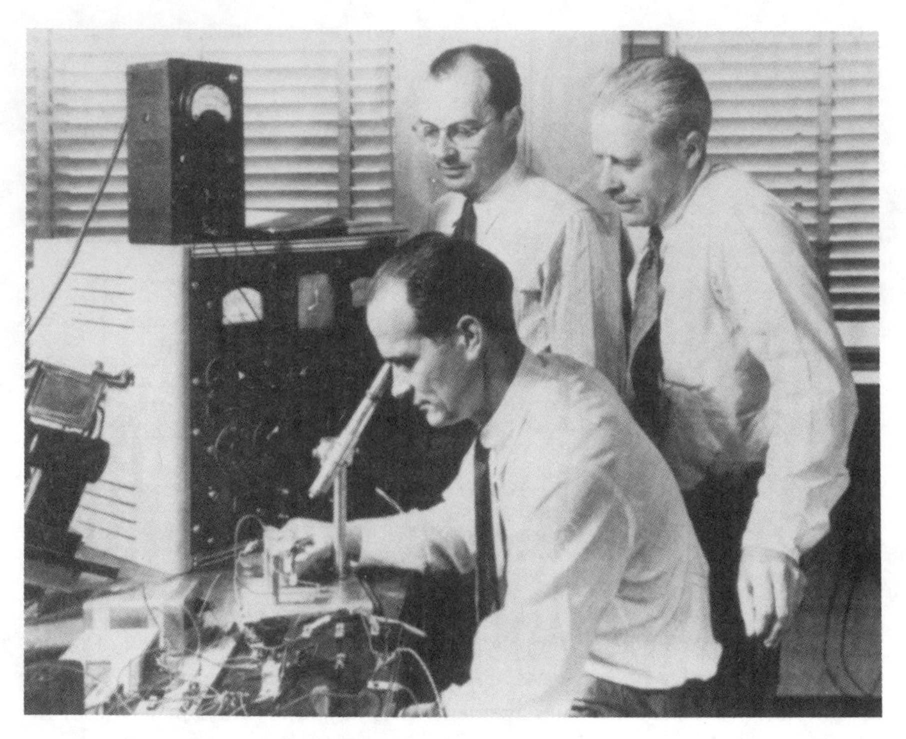

FIG. 55. Bardeen, standing at left, and Brattain observe how Shockley examines the amplifier under a microscope.

outside Bell Laboratories that knew of the invention. The scientists were relieved when they heard that the military had decided not to classify the transistor as "secret." Hence, on July 30 the invention could be shown to the press. The reaction was discouragingly small. The *New York Times* reported it on July 1, 1948, on p. 46 in its regular feature "News of the Radio." First it gave information on entertainment programs. Only at the end was there a short report:

> A new device, called a transistor, with various applications in radio in place of a vacuum tube, was shown yesterday for the first time at the Bell Telephone Laboratories on West 46th Street, where it also was invented [Fig. 56]. The component was shown in a radio which contained none of the usual tubes. It was also demonstrated in a telephone system and in a television system, that was driven by a receiver on a lower floor. In all cases the transistor was used as an amplifier, although it was stated that it also can be used as an oscillator, since it generates and transmits radio waves.
>
> Shaped like a small metal cylinder about 1.5 cm long, the transistor contains no vacuum, grid, plate, or glass envelope to protect it from the air. It operates immediately and requires no warmup time, since no heat is

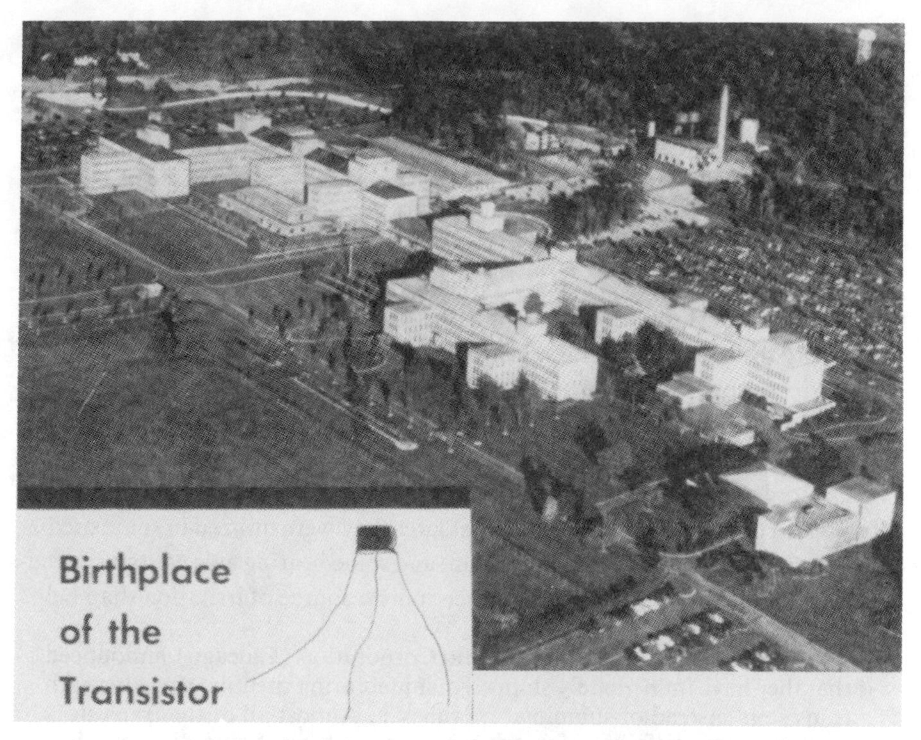

FIG. 56. The "birthday of the transistor." The Bell Telephone Laboratories.

developed as in a vacuum tube. The active parts of the component consist only of two thin wires which run to a piece of semiconducting metal the size of a pin head attached to a metal base. A substance on the metal support amplifies the current introduced through one of the wire tips, and the other wire tip carries the amplified current away.[133]

In spite of the series of accidents which led ultimately to the invention, the transistor was in no way an accidental discovery. The demand for new semiconductive components had resulted in the establishment of a research division devoted specifically to this task. The scientists assigned to this division were those from whom one could expect at the earliest possible time a solution to the problem. The scientists made use of their chance observations in such a way that they led to the achievement of their goal.

The researchers at Bell Laboratories relied on the findings of the Purdue group in their work on the germanium transistor. The Purdue group also observed unexpected behavior by diodes and germanium point contacts (see p. 149ff).

At a meeting of the American Physical Society at the beginning of 1948, shortly after the invention of the transistor, Purdue physicists reported on their experiments with a point contact on germanium. In a subsequent conversation with Brattain, who was also present, one of the Purdue researchers stated that they ultimately would be able to find out what processes were taking place when they attached a second metal wire tip to the germanium surface in order to measure the voltage around the first wire tip. "Yes," answered Brattain, who was still under orders of secrecy, "I think that perhaps that would be a good experiment!" Just such an experiment at Bell Laboratories had led a few weeks before to the invention of the transistor.

Most people saw in the transistor only an alternative to the radio tube, which in the meanwhile had also become a satisfactory component through the work on the radar project, while the transistor still was scarcely more than a laboratory curiosity. Production of point-contact transistors at Western Electric began only in October 1951. A year later they were utilized in some oscillator components in the telephone system, and some hearing aids contained transistors.[17] Indeed, at first they were often more a source of irritation than help:

> At the end of April the Zenith Radio Corporation (Chicago) announced that they have immediately stopped manufacturing their hearing aids with transistors instead of subminiature tubes... . Almost all of the transistors used were unsatisfactory for one reason or another. Some did not withstand shock, and most failed within a few weeks or months because they were not immune to the effects of moisture, which cannot be avoided when the hearing aids are worn very close to the body. It was finally necessary to remove the transistors from all the devices.[52]

FROM THE TRANSISTOR TO THE MICROPROCESSOR

THE TRANSISTOR BECOMES AN INDUSTRIAL PRODUCT

When the first germanium point-contact transistors were produced on an industrial scale in 1951 by Western Electric, the manufacturing arm of the Bell System, several other large firms were experimenting with the new component. It turned out to be difficult to build reliable transistors with the two closely spaced points, and it was almost impossible to produce them with identical characteristics. They had a very limited lifetime and could not process high-frequency currents. Compared with equivalent technical data for amplifier tubes, these transistors were not attractive products. Nevertheless, there was one group that was interested in one particular property of the transistor, its small dimensions. This group was the military, which was well aware of the advantages of miniaturizing electronics.

> A destroyer built in 1937 had only 60 electron tubes on board; in the year 1952 this number had already risen to 3200 (Fig. 57). Since 1950 the Navy had advanced the effort toward miniaturization of electronic circuits through a project called "Tinkertoy" (Fig. 58). Although the transition was already in existence at this time, this project was based on vacuum tubes, and the project was carried out by the National Bureau of Standards. It was overtaken by the new opportunities offered by the transistor, and was closed down in the year 1953, after almost $5 million had been invest-

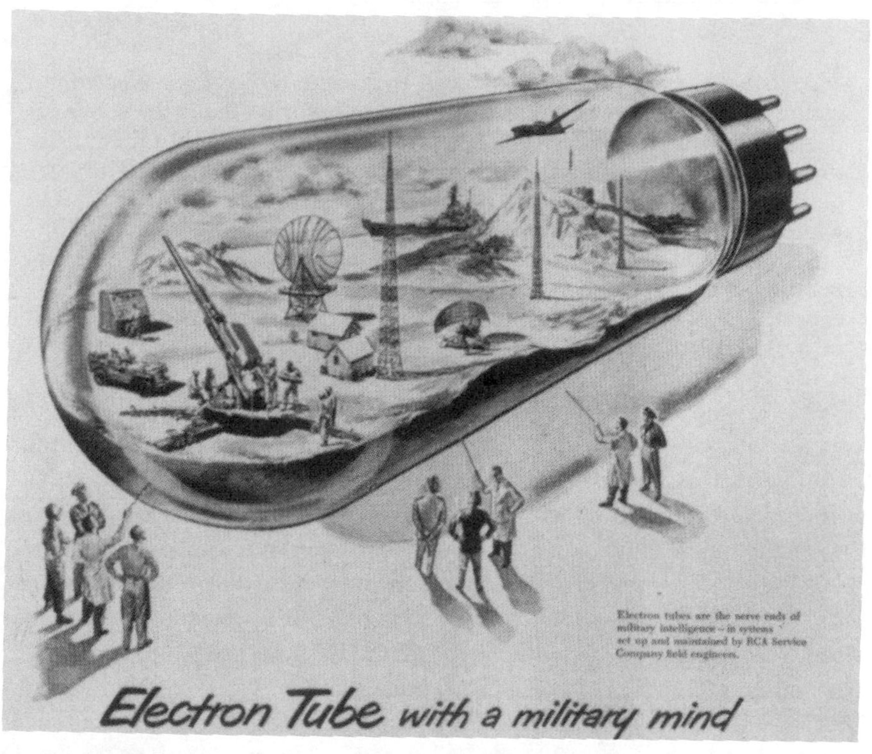

FIG. 57. After World War II the electron tube was used in all types of American weapons. A full-page advertisement in the October issue of Scientific American in 1951 by America's largest manufacturer of electron tubes pointed out the versatile applicability of the "electron tubes with a military mind." "Electron tubes are the nerve ends of military intelligence—in systems set up and maintained by... field engineers." The field engineers explain here to the military the opportunities for using electron tubes.

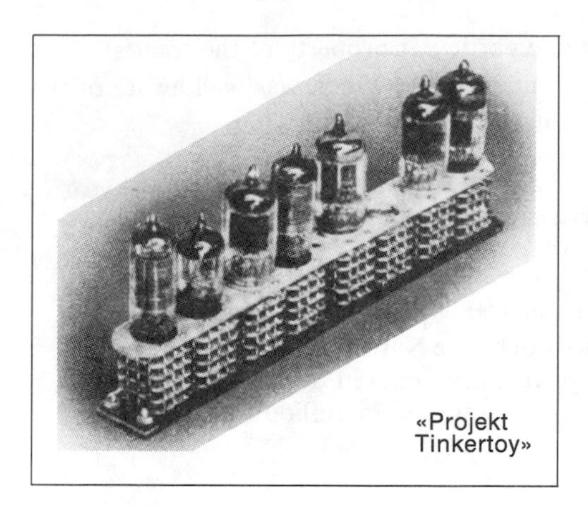

FIG. 58. The Navy's Bureau of Aeronautical Sciences spent $5 million to support Project Tinkertoy to miniaturize electronic circuits. The design of the modules was characteristic of the project. They consisted of several ceramic plates. The project made possible the rapid and inexpensive mass production of electronic devices.

ed. A study in the year 1952 showed that some 60% of the electronic equipment in the Navy did not function satisfactorily. Since the cause of the breakdowns often lay in the mechanical stress placed on the tubes, their replacement by solid-state components appeared very promising. The cost played a subordinate role in all this. A bomb in World War II cost $2500, while in 1952 a new computer-controlled device operating with vacuum tubes was not to be had for less than $250 000. The Air Force considered it possible to transistorize about 40% of its electronic equipment, with an expected savings of 20% in volume, 25% in weight, and 40% (very optimistic) in defects (Ref. 17, p. 228).

In 1951 the Army, Air Force, and Navy assigned to the Signal Corps the responsibility for promoting the development of transistorized equipment for the military. The objective was to make more transistors available, to reduce their cost, and to improve their performance and reliability. Between 1952 and 1964, $50 million of arms money flowed into the American semiconductor industry (Fig. 59), most of it applied to improvement of the production conditions. The military was interested above all in building up adequate production capacity. As a precondition for financial support the firms had to be in a position to promise a production of at least 3000 units per month for every component. The first firms to profit from the military contracts were the large established electronics firms such as Western Electric, the General Electric Company, Raytheon, Sylvania, and RCA. Military funds were expended more abundantly for the purchase of transistors than for research. The transistor manufacturers could count on the military as a reliable customer at prices that would not find takers in the civilian sector (Fig. 60). In 1952 the military purchased almost the

FIG. 59. Between 1952 and 1964 some $50 million of weapons funds poured into the semiconductor industry. More important than the direct support of reseach was the fact that the military become a reliable customer for transistors. The application of transistors was not kept out of advertisement, as shown here in a Tung-Sol ad in Scientific American (1961): "The awesome retaliatory potential of the nuclear-submarine-based Polaris missile describes more vividly than words the vital importance of this weapon in America's protective arsenal." Polaris missiles were equipped with the germanium transistor "Bureau of Navy Weapons type R212."

entire output of 90 000 transistors of Western Electric. Only when the competition for this solid market had led to a drastic lowering of prices were transistors offered in the civilian sector.

In comparing the different transistors debate arose at first about how a good transistor must be manufactured. At the instigation of the Signal Corps, standards for transistor characteristics were developed in collaboration with the Navy, the Air Force, and the principal manufacturers (Ref. 200, p. 76).

The weak points of the point-contact transistor were the closely spaced fine metal wires that pressed against the surface. The spacing had to be maintained exactly, and the arrangement was sensitive to variations in the surface. Just a month after the invention of the point-contact transistor Shockley came up with the idea of burying the transistor effect entirely within the interior of the crystal and replacing the sensitive contacts with adjacent surfaces of differently doped semiconductive materials. He kept this idea to himself while he worked out the details. He had, of course, contributed a great deal to the invention of the point-contact transistor, but he was not present at the critical experiments.

FIG. 60. The first transistors were so expensive that there were no prospects for profitable civilian applications. Even in 1961, when the transistors had won their place after a price drop, General Electric was advertising in Scientific American the low cost of vacuum-tube equipment. The manager wrinkling his forehead asks himself, "Do I have to take into account the high cost of miniaturization?" In the accompanying text General Electric advises him to buy the cheaper tube equipment, even if it is larger.

> Frankly, Bardeen's and Brattain's point-contact transistor provoked conflicting emotions in me. My elation with the group's success was balanced by the frustration of not being one of its inventors. For the next five years, I did my best to try to put Bell Laboratories—and myself—in the lead for transistor patents (Ref. 178, p. 690).

Shockley first revealed his secret in connection with a seminar lecture in which a Bell Labs colleague described an experiment which represented an important step in the direction of Shockley's new transistor. The lecturer had used a transistor design in which the contact points were placed on opposite sides of a thin germanium film, and he demonstrated that holes could move through the interior of the germanium. Now it was too risky for Shockley to keep his secret any longer, and he encouraged Bell researchers to implement his idea. One failure followed on the heels of another. Only when senior researchers at Bell decided to go ahead with the growing of expensive monocrystals was the first junction transistor produced in 1950. When Shockley presented the new component at an international semiconductor conference, it attracted so little attention that it was not even included in the conference report. Further developmental work was necessary to make the junction transistor more attractive. Motivation for this work Shockley received during a visit to the battlefields of the Korean war:

> A group of leading scientists and industrialists...visited Korea recently... . They spent about three weeks in the Far East Command... . Secretary of the Army Frank Pace, in announcing the mission, said that he felt that the effectiveness of the individual soldier can be greatly enhanced by further developments in electronics. The group gathered much valuable information on the problems involved in adapting electronics to the battlefield, so that American technical and industrial know-how may more effectively improve equipment for the Armed Forces (Ref. 7, p. 380).

Shockley, the experienced advisor in strategic research, later reported on his visit to Korea:

> There I learned of the need for proximity fuses that would make mortar shells much more effective. I later discovered a lack of programmes for transistorized fuses and urged that some be started. At Bell Labs, R. L. Wallace stressed that we could do much better if we used not point-contact transistors, but instead nonexistent but possible forms of the junction transistor. This led Morgan Sparks and I to try again, and with Teal's cooperation in early 1951 we made the first microwatt junction transistor...The transistor industry was launched (Ref. 178, p. 691).

The new transistors were more effective and efficient than the point-contact transistors. Their production could be controlled better. A layered structure could be built up during the production of the germanium monocrystals by alternating the doping of the melt with donors and acceptors. The junction transistors, however, still had the disadvantage of limited frequency range.

THE 1952 BELL LABS SYMPOSIUM

AT&T, to which Bell Laboratories and Western Electric belonged, was limited in the use it could make of transistors. A Federal commission kept a watch on the concentration of capital and guaranteed the division of markets and regions. All local telephone companies, for example, had to obtain their equipment from Western Electric in order to assure the technical compatibility of all installations. Western Electric, in turn, was prevented from penetrating other markets, so that they, for example, were not allowed to sell radios, amplifiers, or hearing aids. The firm was a formal partner of all of the patents developed at Bell Laboratories, but these patents could not be exploited in all areas. Bell management in this situation decided on an open information policy in order to obtain as many licensees as possible. In April 1952 Bell Laboratories put on an eight-day symposium on the production of point-contact and junction transistors, on the growing of monocrystals, and on doping. The fee for participation at that time was $25,000. This amount could be applied against later license fees for transistor manufacture (Ref. 160, p. 139ff). The interest of commercial firms was so great that the meeting had to be repeated in order to meet the demand. By the fall 35 license agreements had been signed, nine of them with foreign firms (Ref. 200, p. 75). The meeting had a lasting influence and a strong worldwide growth in the production of transistors began. The Philips firm, which sent three participants to the meeting, had already been working two years on the production of the transistor without success. A few months after the Bell Symposium success was achieved. Toward the end of the same year the scientific laboratory of the Philips concern organized internal colloquia, in order to share the knowledge gained. Philips wanted to begin production as fast as possible in order not to leave the manufacture of this component, which threatened to compete with the vacuum tube, to the "non-tube people." According to one estimate 60% of the 70 million miniature tubes which were installed in military equipment alone in 1951 were replaceable by transistors. And Philips had produced some 3 million of those tubes (Ref. 169, p. 152).

Professionals were well aware that these small electronic components could do much more than duplicate what vacuum tubes were doing. In the Philips annual report for 1952 the importance of the transistor for replacing vacuum tubes was clearly understood:

> Insofar as it is possible to predict the development of semiconductors it must be assumed that the so-called transistor will have a revolutionary effect on the entire field of electronics... . Although one must deal with a partical replacement of electronic tubes, there will be new areas of application which have been closed to vacuum tubes for obvious reasons (Ref. 169, p. 151).

And an article in the American journal *Electronics* reflected on the future role of the transistor:

> As we enter the age of the transistor, it is important that engineers open their eyes open wide to the potentials of these new devices that are like tubes and yet are not like tubes. Circuits can be developed by thinking of transistors as substitutes for tubes. But more important, circuits will come in the harnessing of characteristics that are peculiar to transistors themselves (Ref. 200, p. 75).

NEW FIRMS SPROUT UP

The number of transistor manufacturers increased to 25 during the first half of the 1950s. Six of them began production without previous experience in building electronic amplifying components such as vacuum tubes, for example. One of these firms was Texas Instruments, a manufacturer of geophysical instruments from Dallas. After the firm had obtained a manufacturing license in connection with the Bell Symposium, it established a semiconductor laboratory in 1953 under the direction of Gordon Teal, who had left Bell because he saw better opportunities for implementing his ideas in the new firm. This kind of job hopping was typical among American semiconductor firms from that time on. As a specialist in growing crystals, Teal assisted Texas Instruments in the early successful development of a silicon transistor. Germanium had taken over as the semiconductor material because it was so easy to handle. Silicon is more brittle, chemically more aggressive, and must be heated to above 1400 °C in order to grow monocrystals, but these manufacturing disadvantages are linked with beneficial properties. In contrast to the situation with germanium, the electrons in silicon remain bound in the crystal lattice even at higher temperatures, so that silicon transistors can be used under more extreme conditions. This was a reason for the military to support research in this field. Since silicon transistors also better withstand the heat caused by heavier currents, they are suited for applications involving higher power levels. In addition, this material constitutes a high percentage of the crust of the earth in the form of sand, while germanium is a very rare element. Although the best laboratories, including Bell Laboratories, had been working for years on producing a silicon transistor, the U.S. professional world at a national conference on "electronics in aircaft" in 1954 agreed that another two years would pass before production of electronic components based upon silicon was a reality. But the last paper at the conference, presented by Teal, got the experts excited:

> Many research laboratories of late have become very enthusiastic about studying silicon and its possible use as transistor material. Texas Instruments has made an important step forward... . W. A. Adcock, Morton E. Jones, Jary W. Thornhill, and Edmond D. Jackson had succeeded in producing *npn*-junction transistors from silicon...It so happens that I have one in my coat pocket (Ref. 25, p. 16).

Texas Instruments could be sure of a monopolistic position with this component for three years. The anticipated technical applications of very pure silicon induced laboratories outside of America as well to undertake research in this field. With the development of a new method for growing silicon monocrystals Siemens resumed a significant role in international competition for the first time since the Second World War. In a laboratory which was installed in the Franconian Pretzfeld Castle, several scientists had busied themselves since 1952 with the semiconductor materials germanium and silicon. The process of drawing crystals from the melt, which had been so successful with germanium, failed with silicon because of the higher melting temperature. At this temperature contact with the crucible was enough to contaminate the crystal. In order to avoid any contact with foreign materials, starting in 1953 the crystal was placed across a high-frequency coil and melted a section at a time so that the impurities were gradually moved to one end. Although the researchers at Siemens missed priority on this "contact-free zone melting" by a few weeks (Western Electric scientists were ahead of time), the process turned out to be very lucrative for the firm. Siemens was able to register a series of additional patents on particular steps in the process and eight companies in the United States, Japan, England, Germany, and Switzerland took out licenses based mainly on the know-how developed in Pretzfeld. One of these firms, Wacker Chemie in Burghausen, today is among the leading producers of silicon monocrystals.

Another semiconductor firm appeared in America in 1952: Transitron. It was founded by two brothers, Leo and David Bakalar. Leo was in the plastics business, and David was a solid-state physicist at Bell Laboratories. A point-type contact germanium diode had just been developed there, the gold points of which could handle much higher voltages, and thus it was purchased in large quantities by the military. Transitron offered diodes like this for sale starting in 1953, and with this product rose within a few years to be the second largest semiconductor producer. Opportunity for the new firms came from the fact that semiconductors could be manufactured with relatively small investment in production equipment. With a little capital and a few workers a production facility could be established in a relatively small space. The level of capital required lay within the range that was possible for individual scientists and engineers and groups of them. For this reason the number of scientists among the entrepreneurs rose to levels never seen before. Usually it was also easy to find professional backers for building a semiconductor production facility. Risk capital, usually in amounts that would not have been enough to build a traditional production facility, was injected into this young and growing branch of industry. Above-average profits attracted free capital, but there was always the associated risk that the product manufactured or the method of production could soon become outdated. Those who could not keep up with progress failed. Improved methods of manufacturing were the key to success, and for this reason scientists and engineers with experience in semiconductor production were sought for and paid well. People dissatisfied with decision processes in a large corporation could easily make use of knowledge on semiconductors elsewhere—in their own firm in collaboration with colleagues or in a small company where they could play a dominant role.

The successes of the new firms were spectacular. Though most of the research work came from the large electronic firms, these large firms could not maintain their market position in the competitive environment. In 1957 the old firms already had been reduced to 36% of the semiconductor market (Table VIII).

SILICON VALLEY

In 1954 William Shockley left Bell Laboratories to establish his own firm. He chose as its site his home town of Palo Alto, California, where conditions were good for setting up a modern industrial enterprise. The railroad magnate Leland Stanford had established a private university there in memory of his prematurely deceased son Leland, Jr. Today this is one of the elite universities of America. Even before the war Fredrick Terman, an influential professor of electrical engineering, attempted to combine at Stanford the needs of university research with the demands of a scientifically based branch of industry. He encouraged his students to establish independent enterprises in the vicinity of the university. Among those who followed his advice were David Packard and William Hewlett, who began to build electronic equipment in Packard's garage. Today their names stand for one of the largest manufacturers of electronic measuring equipment.

Shockley established the Shockley Semiconductor Laboratory in the same area. His scientific reputation attracted large talent to Palo Alto, among them Gordon E. Moore and Robert Noyce, both from established electronics firms in the East. While Shockley was still working for Bell Labs, he had invented a new

TABLE VIII. The market shares of old and new firms in the American semiconductor market in 1957. Western Electric and the vacuum-tube manufacturers, which had begun the production of transistors, were increasingly being forced from the market by new ambitious firms.

Western Electric		5%
Vacuum-tube manufacturers		
General Electric	9%	
RCA	6%	
Raytheon	5%	
Sylvania	4%	
Philco–Ford	3%	
Westinghouse	2%	
Others	2%	
Subtotal		31%
New Firms		
Texas Instruments	20%	
Transitron	12%	
Hughes	11%	
Others	21%	
Subtotal		64%
Grand total		100%

electronic component. With two connections it was a diode, while its three *pn* pathways endowed it with bistable characteristics. The two states could be used to store information. This Shockley diode would find application above all in the telephone business and it was the first product of the new firm (Fig. 61). The silicon technology turned out to be disproportionately more difficult than the germanium technology, and for this reason the yield of useful diodes was very small. Shockley undertook research to improve the output, but the cost of these efforts was too much for this small firm. The young colleagues strove to apply the experience already gained to manufacture transistors. Significant success had already been achieved in secret experiments and they did not need to worry about a market. Military and space organizations paid the highest prices for reliable transistors which were not sensitive to temperature fluctuations. But Shockley refused to give up his larger plans for production of silicon transistors. Eight colleagues led by William Noyce realized the opportunities available to them, and left the firm. Noyce went to an investment office and was referred to the Fairchild Camera and Instruments Company, which was interested in semiconductor engineering in the field of aircraft building. The company manage-

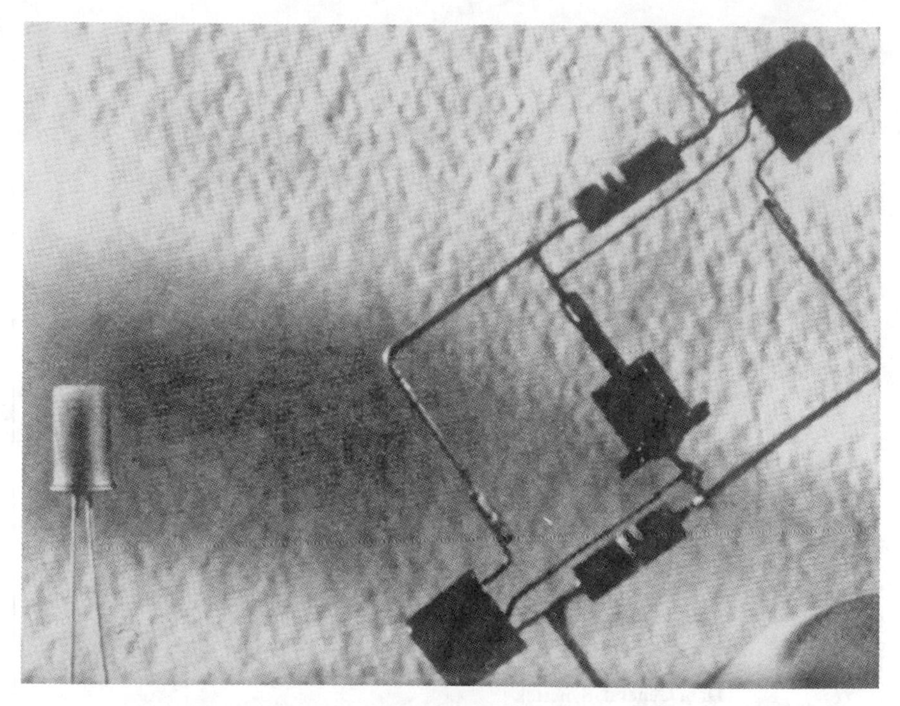

FIG. 61. Shockley's four-layer diode on the left does the same job as the five-component circuit on the right.

ment agreed to finance an independent semiconductor firm, but reserved the right to absorb the enterprise entirely into the Fairchild Company after an interval of two years.

In the fall of 1957 the "eight traitors" as they were called by Shockley, set up shop a few miles from Palo Alto in Mountain View. Through additional splits and new mergers during the next 20 years about 100 firms were created (Table IX), all of which took root in the general vicinity of Stanford University. Technical talent, for the most part, remained in this vicinity in order to find new positions. Job hopping was the order of the day. This led to a scientific and technical clique situation in which every scientist knew everyone else and practically everyone had worked at one time or another for or with everyone else. The Santa Clara region, which was famous for its orange plantations, where this industrial explosion took place became "Silicon Valley," symbol of a new era.

Shockley attempted to save his firm by emphasizing more strongly the production side of things. But success eluded Shockley's transistor laboratory

1955	Shockley Transistor, Clevite, ITT
1956	
1957	Fairchild Semiconductor
1958	
1959	National Semiconductor, Rheem Semiconductor
1960	
1961	Signetics, Amelco, Raytheon Semiconductor, H.P. Associates
1962	Siliconix, Molectro
1963	Stewart Warner Microcircuits, General Microelectronics
1964	Union Carbide Electronics
1965	
1966	Philco-Ford Microelectronics, American Micro-Systems, Cal-Dak
1967	National Semiconductor, Electronic Arrays, Intersil
1968	Cermetek, Monsanto Electronics, Avantek, Lab-Go, Integrated Systems Technology, Nortec, Kinetic Technologie, Intel, Computer Micro-Technology, Qualidyne, Electro Nuclear Labs, Advanced Memory Systems, Precision Monolithics,
1969	Lithic Systems, Communications Transistor Corp., Monolithic Memories, Cartesian, Advanced LSI Systems, Signetics Memory Systems, Advanced Micro Devices, Four Phase
1970	Litronix, Integrated Electronics, Varadyne, International Computer Modules
1971	Cal-Tex, Exan, Micro Power, Intersil Memory, Standard Microsystems, Antex
1972	LSI-Systems, Nitron, Frontier Electronics, Interdesign, Light Emitting Devices, IC Transducers, Opto Ray, Optical Diodes,
1973	Data General, Synertek
1974	Monosil, Zilog
1975	Mnemonics, Maruman Integrated Circuits, Exonix, Semi Processes
1976	Supertex, Cognition, Integrated Technology Corp.

TABLE IX. New semiconductor firms in Silicon Valley, 1955–1976. Fairchild provided at least one founder for 24 firms.

and in 1959 the firm was purchased by Clevite. The latter firm in turn was taken over in 1966 by ITT.

Fairchild began production of silicon transistors using the most modern technical methods. A new process for manufacturing *pn* junctions was developed at Bell Laboratories, and this was presented to the rest of the industry at a second symposium in 1956. The foreign atoms (donors or acceptors) were introduced into the semiconductor by wide area diffusion. The area necessary for the component was uncovered by means of a photoetching technique. The thickness of the base could be precisely controlled to a few thousandths of a millimeter by diffusion, ten times better than with earlier methods. Because of their resemblance to a mesa, these transistors were called mesa transistors. It was possible with this technology to put several transistors on one wafer, which was then split up to wire the transistors to several electronic circuits. Mesa transistors were sturdier than their predecessors, they withstood higher currents and were "fast," that is, they could operate at speeds up to a billion cycles per second. In spite of the price of up to $100 each, they became an important sales item for Fairchild. The sputnik shot of 1957 contributed to the success of the mesa transistors. A rocket had propelled an artificial satellite into space for the first time, but the satellite had one blemish in that it had been developed by Russian scientists. American military leaders were alarmed. Indeed both superpowers had atom bombs, but appropriate means of delivery were still lacking and here the East appeared to have achieved a leap forward. In 1958 the Army and the Signal Corps pulled the dormant Project Tinkertoy (see p. 168) from the drawer and supported the development of micromodules, this time based on semiconductors. The Navy favored the concept of building micromodules from thin films and starting in 1958 financed a research and development program in this area. The Air Force, which stood to profit most from the use of miniaturized circuits, tried to think as far ahead as possible beyond the transistor. If the functioning of the transistor involved a control of electrons in semiconductive layers, then it should also be possible to build a type of circuit in which this control would be spread over the entire crystal. In this way the idea arose of developing an integrated component for use in solid-state electronics (Fig. 62).

Fairchild's mesa transistor outperformed all other types. Through the diffusion technique the active layers, however, were moved closer to the surface and thus were susceptible to moisture and dirt. In order to protect the surface Fairchild engineers applied a discovery which had been made by researchers at Siemens and Western Electric. A thin oxide layer prevented the inward diffusion of doping materials and thus could shield the *pn* junction from external influences. If the protective film contained particular foreign atoms, it was especially effective. Germanium, which up to that time was the most heavily studied semiconductor after silicon, could not be oxided to form a protective layer. This was a major factor responsible for the increasing importance of silicon in further developmental work. The silicon crystal could be treated from the surface. Openings were etched in the oxide layer through which the necessary foreign atoms could be diffused. The silicon oxide protective layer served at the same time as insulation. This new technology produced transistors whose functioning was restricted to a thin surface layer. Hence it was given the name planar technology.

FIG. 62. Power and microcomponents. "Since the invention of the tiny transistor 12 years ago, defense systems have required a continuously higher reliability, and lower costs." Like many other companies Transitron heeded this message from the military, which demanded stronger miniaturization projects following the shock of Sputnik. In 1961 Transitron presented its latest microcomponents in a Scientific American advertisement, components which were ten times smaller than their predecessors. "...now available and already used in missile weapons and satellites, these tiny semiconductors contribute every day to the increased success of our missile and space travel program."

THE INTEGRATED CIRCUIT

Many scientists considered the production of transistors to be uneconomical. The new procedures made it possible to produce many transistors on a single wafer of silicon, which then was cut apart to link the transistors again into circuitry. One of these scientists was Jack Kilby, who had begun his career as an electronic specialist at Globe-Union's Centrallab. Kilby had participated in the first Bell Laboratories Symposium and was fascinated by the invention of the transistor. He led the development of small transistorized hearing aids, but was unhappy with the conditions surrounding the semiconductor development work at Centrallab. In the spring of 1958 Kilby applied for positions at several firms, emphasizing his interest in miniaturized circuits. Texas Instruments, which had signed a contract in connection with the micromodule program of the Signal Corps, hired Kilby. Kilby later reminisced about his early days with Texas Instruments:

> At that time Texas Instruments had a company vacation, that is the company simply closed down during the first weeks of July and everyone with vacation time coming took it at this time. Since I had just started there and did not yet have any vacation time coming, I stayed quite alone in the empty facility... . I began to look around for alternatives, and the idea of the integrated circuit actually appeared during this fourteen-day company shutdown (Ref. 68, p. 122).

Kilby knew that more than transistors and diodes could be made from semiconductor material. The material could also be used to form ohmic resistors and condensers (by opposed hookup of *pn* junctions). He sketched a simple circuit, an oscillator, that was made up entirely of semiconductive components (Fig. 63).

Kilby's idea aroused great interest among the management of the firm. In order to demonstrate that such a circuit would function, he was asked first to build it from individual semiconductive components. Kilby used the latest components from the Texas Instruments production line, germanium mesa transistors, 26 of which could fit on one wafer. He cut a germanium wafer into appropriately sized pieces and connected the components with fine gold wires. This first integrated circuit (IC) contained five electronic components: three resistors, one condensor, and a transistor (Fig. 64).

In February 1959 Kilby applied for a patent on miniaturized electronic circuits. Even though in the actual model the individual parts had to be connected with wires, the application already contained the following sentence: "...Then electrically conductive material such as gold can be melted on the insulating material in order to make the necessary electrical circuit connections" (Ref. 68, p. 124).

Texas Instruments announced the development of a solid-state circuit "no bigger than the head of a match" with a big advertising campaign and sold them for $450 each.

The news about the development of the integrated circuit speeded up the work at Fairchild, where Noyce had already independently been looking for the best way to attach several components to silicon wafers during the manufacturing process by conductor pathways. In the planar technology Fairchild had the most suitable manufacturing methods. Noyce described its advantages:

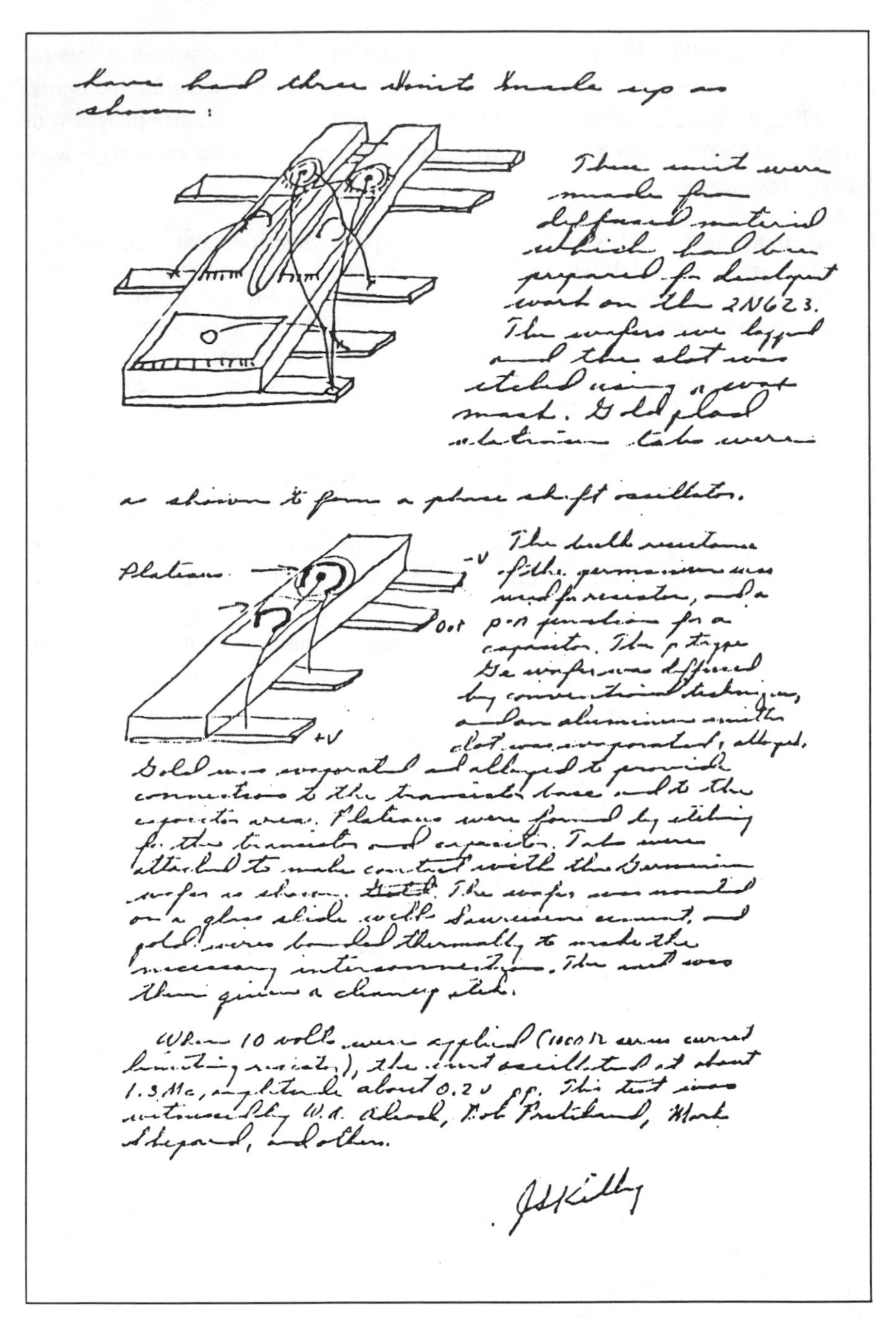

have had three units made up as shown:

These units were made from diffused material which had been prepared for development work on the 2N623. The wafers were lapped and the slot was etched using a wax mask. Gold plated platinum tabs were

as shown to form a phase shift oscillator.

The bulk resistance of the germanium was used for a resistor, and a p-n junction for a capacitor. The p type Ge wafer was diffused by conventional techniques, and an aluminum emitter dot was evaporated, alloyed. Gold was evaporated and alloyed to provide connections to the transistor base and to the capacitor area. Plateaus were formed by etching for the transistor and capacitor. Tabs were attached to make contact with the Germanium wafer as shown. ~~Gold~~ The wafer was mounted on a glass slide with Sauereisen cement, and gold wires bonded thermally to make the necessary interconnections. The unit was then given a cleanup etch.

When 10 volts were applied (1000 Ω series current limiting resistor), the unit oscillated at about 1.3 Mc, amplitude about 0.2 v p.p. This test was witnessed by W. A. Adcock, Bob Pritchard, Mark Shepherd, and others.

J S Kilby

FIG. 63. Jack Kilby's design sketches for a simple integrated circuit.

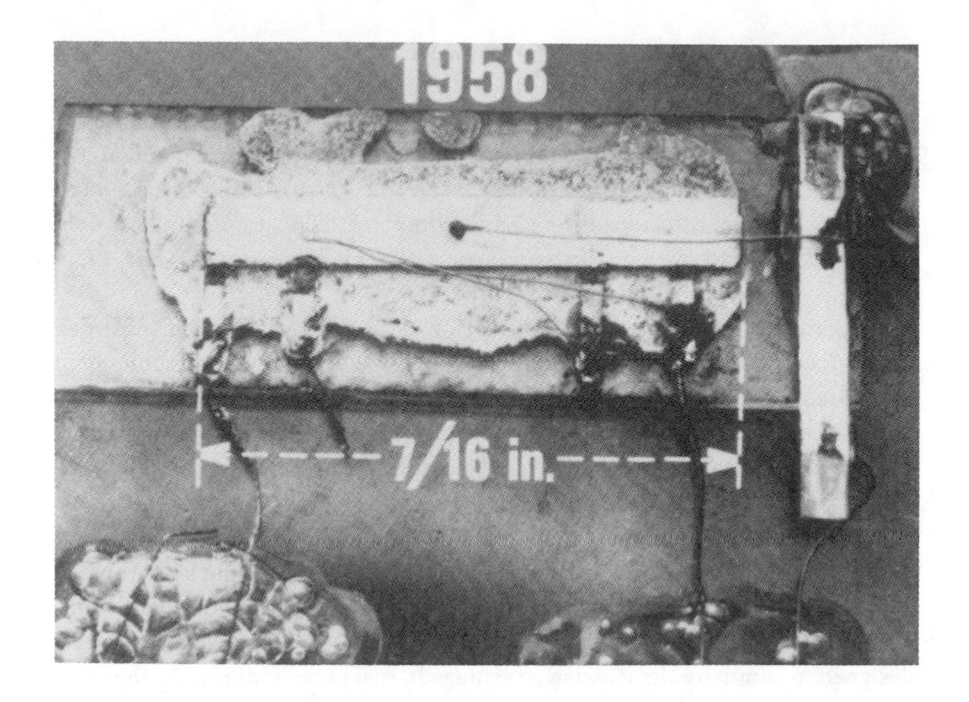

FIG. 64. The first integrated circuit, built by Texas Instruments on the basis of Kilby's sketch. ($\frac{7}{16}$ in. equals about 11 mm).

"In planar [the reference is to the corresponding technology] a lot of transistors and other circuit elements were imbedded in silicon, and these circuits could be electrically insulated by the insulating layer instead of cutting them apart and wiring the pieces together again" (Ref. 68, p. 124).

In July 1959 Noyce also applied for a patent on an integrated semiconductor circuit with "Layered...metal strips extending beyond the insulating oxide layer and lying upon it, in order to produce electrical connections to and between various areas of the semiconductor material" (Ref. 68, p. 125).

In 1962 Texas Instruments complained of patent infringement. A vehement controversy developed concerning the invention and the design. It was settled only in 1969, when in the final instance the U.S. Board of Customs and Patent Appeals awarded ownership of the invention to the two scientists for the different ground that was covered: to Kilby for the basic invention, and to Noyce for the modified structuring.[25]

As was the case with the transistor, the public at first greeted the integrated circuit with skepticism. As long as the failure rate for transistors remained at several percent, it seemed pointless to combine several functioning transistors on a single chip. But the main stumbling block was the high cost of manufacturing. The following statement concerning an early Texas Instruments circuit has been ascribed to a director of the Philips company: "This thing replaces only two transistors and three resistors and costs $100. They are crazy!" (Ref. 17, p. 98).

Again as in the case of the transistor, it was the military, which disregarded the high costs, who were the first customers. Table X shows the percentage of integrated circuit purchases which were connected with the military. In 1960 Texas Instruments announced its first commercial IC; it was ordered by the military. A year later the firm delivered a complete computer to the Air Force that weighed only 280 g and contained 587 components. The equipment that was replaced had 14 times as many parts and was 150 times larger. In 1962, Texas Instruments received a contract from the military to design a "family" of 22 special circuits for Minuteman rockets. This program to develop an intercontinental ballistic missile was a part of the expansion of space and defense projects designed to eliminate the Russian advantage in rocket technology. At the same time Fairchild also received contracts from the National Aeronautics and Space Administration (NASA) to develop integrated circuits for rockets and satellites. In 1961 the firm put its first integrated circuit on the market, the circuit replacing four conventional transistors and two resistors.

Year	Average Price	Military Portion
1962	$50.00	100%
1963	31.60	94
1964	18.50	85
1965	8.33	72
1966	5.05	53
1967	3.32	43
1968	2.33	37

TABLE X. Average price of integrated circuits and the portion of the total integrated circuit production purchased by the military. As shown here, in 1962 the military purchased all the ICs produced, and in 1966, when the price had already fallen to one-tenth of its original level, the military was still buying more than half the total produced. By 1972 the price had dropped to about $1.

MOS TECHNOLOGY

The transistors used in the early integrated circuits were based on the same principle exemplified in the first transistor. Since current was carried by both holes and electrons, this class of transistors was called "bipolar." When Bardeen, Brattain, and Shockley attempted to build a solid-state amplifier, they searched first for a simple circuit design that would involve one kind of charge carrier. The experimental testing of this unipolar transistor, however, failed because of the poor quality of the surface. The influence of an electric field, which was to be used to control the flow of charge carriers, was nullified by irregularities in the surface. During a decade of transistor development and production the engineers learned how to handle the semiconductor material and its surface. The bipolar transistors themselves, since the adoption of the planar technology, consisted of thin layers on the semiconductor surface. Scientists at Bell and RCA set themselves the task of putting the old idea into practice using newer technology so that from the beginning many transistors would be connected together on a single chip. The principal problem was to protect the circuits against contamination during production.

By 1962 the first commercial field-effect transistors had been produced (see Appendix, p. 215), following the invention of the simplest type of transistor by Steven Hofstein and Fredric P. Heimann, two young engineers at RCA; two doped regions on the silicon surface were covered with a layer of silicon oxide and a metal film was vapor deposited on top. The arrangement was called an MOS transistor, to indicate the layers used (metal-oxide-semiconductor). By means of the metallic film, the gate, current flow could be controlled between the doped source and drain regions. Depending on the polarity the semiconductor could be enriched with or depleted of charge carriers, so that the current flow was controlled.

Because of its simple design, the MOS transistor was particularly well suited for the production of integrated circuits. Hofstein and Heimann demonstrated this when at the end of 1962 they squeezed 16 MOS transistors on a 2.5 square millimeter chip to make a circuit with many uses.

The number of circuit elements crowded on a single chip doubled, on the average, every year. Starting from a few individual components in the year of 1960, the number had risen to 100 000 in 1975 (Figs. 65–68). At the same time, the price of individual transistors fell. For example, in 1964 Texas Instruments was selling 100 junction transistors packed in plastic for only 75¢. Based on the cost of today's chip, 100 transistors cost less than a penny.

Beyond its advantages of higher packing density, smaller power consumption, and a simpler manufacture the MOS transistors also had disadvantages.

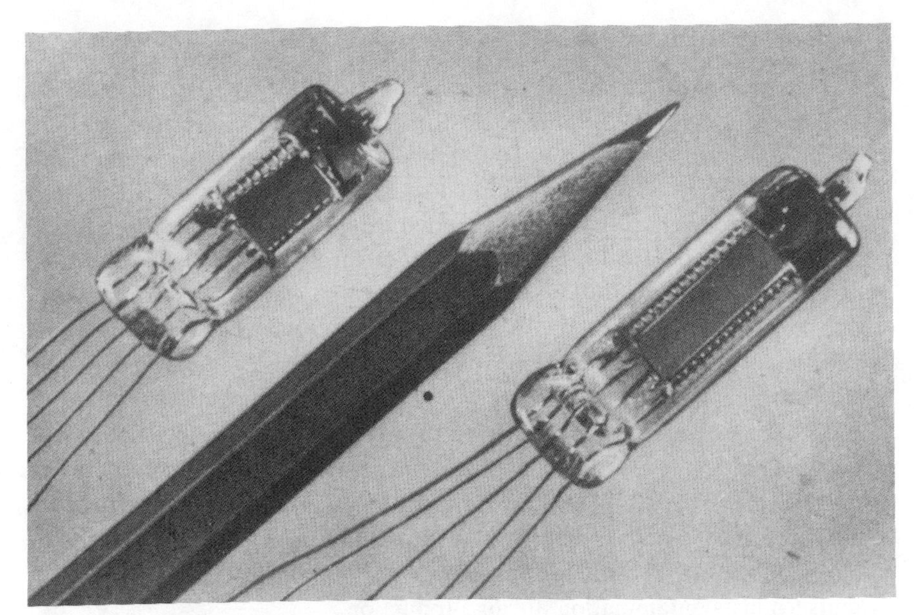

FIG. 65. The effort to miniaturize electronic components is not an invention of the transistor age. Ever since the military began to use electronic equipment for its own purposes, savings in size and weight have been paramount. The size comparisons used to demonstrate progress in miniaturization testify to the rapid development in this area. A comparison between a pencil and a vacuum tube in 1948.

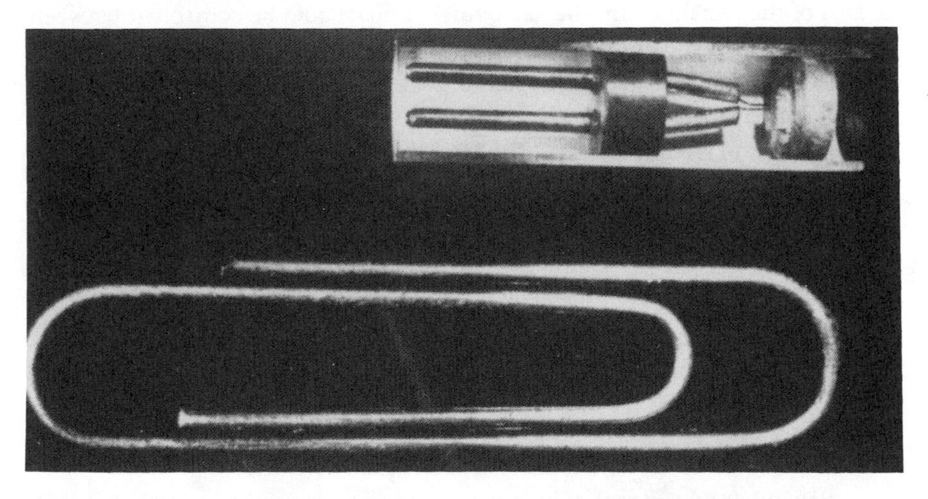

FIG. 66. The transistor, which performs tasks similar to those performed by vacuum tubes, is compared with a paper clip in 1948.

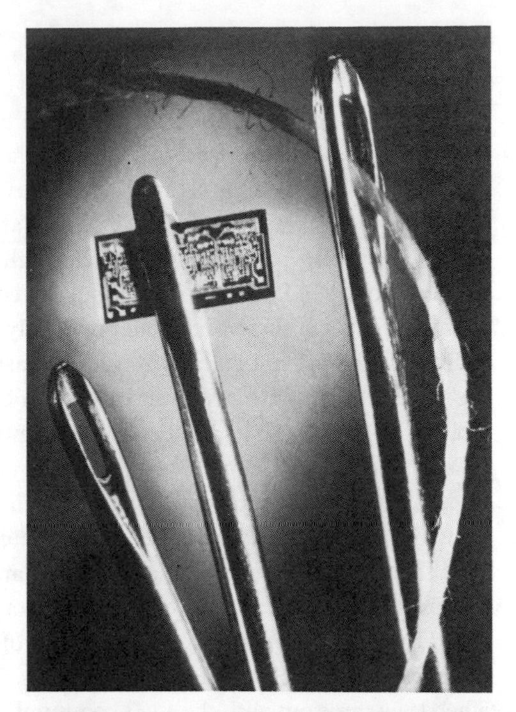

FIG. 67. Scarcely 20 years later it was already possible to pass 120 circuit elements through the eye of a needle.

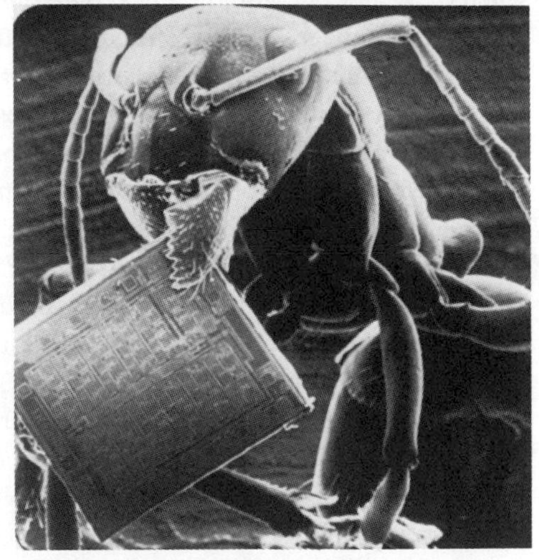

FIG. 68. After another 15 years a tiny insect resembles a monster next to a clip containing ten of thousands of components.

They were extremely sensitive to electrostatic loading. The tiniest overvoltage could break through the thin oxide layer and destroy the MOS. In addition, they operated significantly slower than bipolar transistors. But the MOS were the cheapest form of integrated circuits and thus the most suitable for the consumer market.

MICROPROCESSORS

With the appearance of the integrated circuit the American electronic industry was split even more clearly into the two camps of traditional firms and new companies, which often won a large share of the market with production of a single product. The number of companies in Silicon Valley continued to increase: three new manufacturers of integrated circuits appeared in 1966, another three in 1967, thirteen in 1968, and eight in 1969. Their names consisted of permutations of a few basic syllables like Tech, Inter-, Micro-, -Ics, -Tron, etc. Many founders of these firms had previously been scientists and engineers with Fairchild who did not agree with the increasing restrictions placed on them by the mother firm, Fairchild Camera. Morale was bad even in the front office. Robert Noyce, who had prospects of becoming President, and Gordon Moore, Director of Research and Development, quit the company to concentrate on integrated circuits for computer storage in a new company. Together with Andrew Grove they founded INTEL (Integrated Electronics) and set up shop on land that was formerly a pear orchard in Santa Clara.

In the 1950s and 1960s electronic storage components in computers consisted of arrangements of cores and rings of magnetic material about 1 in. in diameter in a network of thousands of wires. INTEL attracted attention in 1970 when they came out with the 1103 computer storage chip. The ones and zeros of computer language could be stored as electrons present or absent in miniature capacitors which were charged and discharged through transistors. About 3000 transistors were integrated in an area of less than ten square millimeters, and these could be used to write, store, change, or retrieve 1024 units of information. Compared with core storage, the RAM (random access memory) storage chips had a substantially smaller access time while using less power and costing considerably less. They spelled the end of the magnetic storage industry (Ref. 68, p. 167). While the storage technique was still under development INTEL received a contract from a Japanese desk calculator firm, Busicom, to develop a series of chips for a new generation of calculators. Busicom provided a preliminary design and a design team. The development project was placed under the supervision of Ted Hoff, who was Professor of Chemistry at Stanford University and had signed a consulting contract with INTEL. Instead of designing several different chips for different models of the desk calculator, Hoff suggested that a programmable chip be developed which could be adapted to the different requirements. INTEL management allowed Hoff to follow up on his idea, while the Busicom team continued to design a variety of special-purpose chips. Hoff integrated all arithmetic and logical processing elements on a single chip, a total of about 2250 circuit elements. This meant that all of the functions of the central processing unit (CPU) of conventional computers were contained in an area measuring 3×4 mm. When this chip was combined with others containing storage and additional registers, and provided with input and output units, it could be operated as a microcomputer. INTEL called the chip a microprocessor.

INTEL at first did not realize the wide-ranging consequences of this achievement. After Busicom had approved the concept, the two firms signed a contract that required INTEL to deliver the new components to Busicom exclusively. Only as the development work progressed did INTEL see more clearly what possibilities were closed off by this contract. INTEL management then used Busicom's financial difficulties to renegotiate the contract. The companies agreed that Busicom would get the chip for a lower price and in exchange INTEL could market it.

The first microprocessor was offered for sale in November 1971. "The actual invention of the microprocessor," said Hoff later, "was not as significant as recognition that there was a market for something like this" (Ref. 68, p. 148). The next application of the microprocessor was in the microcomputer. Just a few chips in a housing could accomplish more computing and storage than the large installations of the 1950s, which filled entire rooms (Figs. 69 and 70). In 1975 a microcomputer could accomplish more than 100 000 logical operations

FIG. 69. Eniac, the first program-controlled electronic computer from the year 1945, weighed about 30 tons, contained 18 000 electron tubes, and a total of about 500 000 components.

FIG. 70. Today's microprocessors, like the IMSAI 8084, are far superior to the large computers like Eniac in computing speed and storage capacity.

in a second. But the market for microcomputers appeared very limited. The entire demand was estimated to be about 20 000 machines per year, and with an expected market share of about 10–20% this amounted to a few thousand chips per year, not an especially attractive business prospect (Ref. 68, p. 150).

A further drawback was the fact that INTEL had no experience with the writing of computer software. Nevertheless, the company decided to sell both the microprocessor and a microcomputer, the MCS-4. Just six months later a new model of the microprocessor came on the market with twice the performance of its predecessor, and by 1974 there were 19 different microprocessors offered by various firms. The success of the microprocessor lay in the fact that not only was it able to uncover the existing market for small computers, but it opened numerous new markets, just as the transistor was far more than simply a replacement for vacuum tubes.

Some of the first INTEL microprocessors were purchased by a small electronics firm in Monterey and incorporated into a computer/digital watch. Then the storage chips were arranged somewhat differently, attached to a loudspeaker, and sold as a microcomputer phonograph that could play tunes. Microprocessors were also used to control industrial processes. Where previously a mainframe computer processed the measurements of numerous testing devices and called for the necessary adjustment, microprocessors were built into every test device and the central computer was dispensed with. Automation was thereby greatly reduced in cost. Microprocessors were also incorporated into vending machines, cash registers, and traffic lights.

The number of manufacturers also increased rapidly. Advanced Micro Devices followed INTEL; then Fairchild, Signetics, American Micro Systems, and other firms in Silicon Valley entered the microprocessor market. Among the traditional electronics giants there were Rockwell, RCA, and aerospace firms. Soon Japanese firms appeared, and Siemens and Philips produced microprocessors as well. The prices of the processors and the associated circuits dropped. The chip manufacturers estimated a major part of the market wrongly at the beginning. Noyce conceded:

> The entire consumer market was an area which we had originally not envisioned at all. It seemed simply impossible that this phenomenal step in electronic miniaturization represented by the microprocessor would ever become so inexpensive that the simple consumer requirements could be satisfied. Take door locks as an example. Today there are many microprocessor-controlled door locks, but at that time it appeared simply impossible to reduce the costs of this very modern electronic device sufficiently to compete with the price of an ordinary lock. The home and hobby microcomputer market was also an area which we did not see at the beginning, as was the entire field of electronic games (Ref. 68, p. 157).

Microprocessors were incorporated into electronic organs, into color television sets, into stereo sets, food-dispensing machines, microwave ovens, telephone systems, automobiles, and so on.

The 1970s in the jargon of the field were the years of large-scale integration (LSI), that is, large integrated circuits with a few thousand elements on a single chip. The years of very large-scale integration (VLSI) began with the 1980s, when we saw storage chips with room for 100 000 active elements. The size of the individual elements is measured in a few thousandth of a millimeter. The zero-one information is represented by electrical charges which are so small that previously unknown errors crop up: elementary particles from the cosmic rays can change individual bits in the storage elements, producing the so-called "soft" computer errors. Thus there appear to be physical limits to the extent to which storage capacity or computing speed can be increased by miniaturization.

Scientists in the laboratories of electronic firms are trying to find new ways to break through the miniaturization barriers. Completely different properties of solids appear to have promise for increasing the storage capacity of chips and the computing speed of computers. One such property is the superconductivity of many metals at low temperatures (see p. 56ff). Using this property the world's fastest circuit component has been built, called the Josephson junction, after its inventor. It consists of two tiny metallic films separated by a thin insulating layer. When cooled down to superconducting temperatures, that is, a few degrees above absolute zero, it possesses two switching positions like the transistor. Only one ten-thousandth of the energy of a transistor is required to operate it, and its amazing switching speed points toward computer performance so far unimagined. Such a supercomputer would be scarcely larger than a tennis ball and could perform up to a billion binary operations (zero-one decisions) in a second.

Another possibility being pursued by the computer visionary is the "biochip." Its smallest circuit components will be biological molecules, which regulate, for example, the electrical charge transport through the membranes of biological cells. Measuring only a few millionths of millimeter, such components are a thousand times smaller than circuit elements of today's ICs. The military has shown strong interest in the research going on in this area, which involved biologists, chemists, physicists, electronic experts, and computer professionals working together. The Second International Conference on Molecular Electronics was held at the U.S. Naval Research Laboratory in Washington in the spring of 1983. The principle of a diode based on this technology has already been successfully demonstrated and patented by IBM researchers in partial steps. An organic molecule was made to act as an electron donor and another as an electron acceptor. Both parts should be joined by an insulated molecular grid, which allows electrons to pass in only one direction when a voltage is applied. Just like natural proteins in living organisms, computer molecules could also be produced by artificially created enzymes so that a self-reproducing machine can be conceived.[163]

In order to develop faster electronic components, in recent years substances have come into use whose semiconductive properties had been found as early as three years after the invention of the transistor, but for a long time they were hardly utilized by industry. In 1951, Heinrich Welker applied for a patent entitled "Electrical Semiconductor Device" with the purpose of imitating available semiconductors in group IV of the periodic system of the chemical elements. As leader of the Solid State Research Division of the General Laboratory established by the Siemens firm in Erlangen after the war, he pursued a program of work that investigated compounds like indium antimonide and gallium arsenide instead of germanium and silicon as semiconductive substances. As Welker had figured in advance, the new substances exceeded the traditional materials in mobility of electrons and holes—as much as several times when in the form of pure monocrystals.[202] The new materials, however, also appeared to have disadvantages as well. The properties had been studied in far less detail than in the case of silicon, and there was no experience at all in producing and working them. Under the slogan "better one semiconductor for everything, even if it is not optimally suited everywhere" the electronic firms chose to stick to the traditional silicon. The III–V compounds only became of interest again when the physical limits of material properties had been reached.

The high electron mobility of gallium arsenide was utilized in novel components which cropped up in the literature with names like "high speed logic" and "super high speed computer applications." Today thousands of circuit elements can be placed on a single chip of this material.

Another property of the III–V compounds, their wide forbidden zone, finds application in optical electronics, a new field of electronics involving products like light-emitting diodes and semiconductor lasers.

In contrast to the producers of the first transistors and integrated circuits, the microchip producer no longer had to resort to military clients. The production processes for manufacturing chips had been developed so far that the products could be offered at a price that was also within the means of civilian users. The high growth rate enticed many investors into the field of microelectronics. In 1981 the highest growth rates for electronic devices were seen in text processing systems and in the armament field. Table XI gives an overview of the growth rates in the most important fields of application (Ref. 48, p. 128). Competition among the firms increased the rate of microelectronic progress so that companies were driven to feverish innovation and still are. Components and products become obsolete faster than in any other field, and only firms which achieve an advantage over the competition through attractive products or cheap production processes can be sure of making high profits for a couple of years. After one or two years imitators appear and prices and profits drop. Only a firm which is at the forefront of innovation with a product and is ahead of its competition can reach a level of profit required for reinvestment in research and development.

The expenditures on research and development also changed the employment structure in the semiconductor industry (Table XII). The demand for scientists, engineers, and technicians increased most strongly while the number

TABLE XI. Growth rate for microelectronic devices in 1981. Text processing systems showed the fastest growth rate (30%), followed closely by electronic equipment for armament (28%).

	PERCENT PER YEAR
Data processing systems	8
Data storage systems	12
Peripherals	18
Office copiers	25
Text processing equipment	30
Office and other equipment	15
Transmission sector:	
Telecommunication	24
Radio/television	14
Data transmission	11
Industrial control equipment	18
Text and measurement equipment	14
Automobiles	22
Medical equipment	15
Other regulating equipment	17
Audio equipment for consumers	4
Household equipment	6
Devices for personal use	8
Videorecorders	9
Games, etc.	14
Television	6
Weapons	28

of blue collar specialists declined drastically (Ref. 48, Fig. 185). The increase in the proportion of solid-state physicists among the scientific staffs increased quite noticeably when miniaturization of the elements in the most highly integrated circuits bumped against physical barriers. Their areas of expertise included not only the development of alternatives to circumvent the limits of miniaturization, but also the improvement of input and output devices for communication with the computer. Previously, simple switches or keys were used for the input of data. In the near future it will be possible to use the human voice for this purpose. In areas where data input is not by humans, but rather by industrial robots, sensors will transmit information on temperature, location, orientation, speed, electromagnetic fields, humidity, chemical concentration, etc., in real time to microcomputers which control the production process (Ref. 82, p. 268).

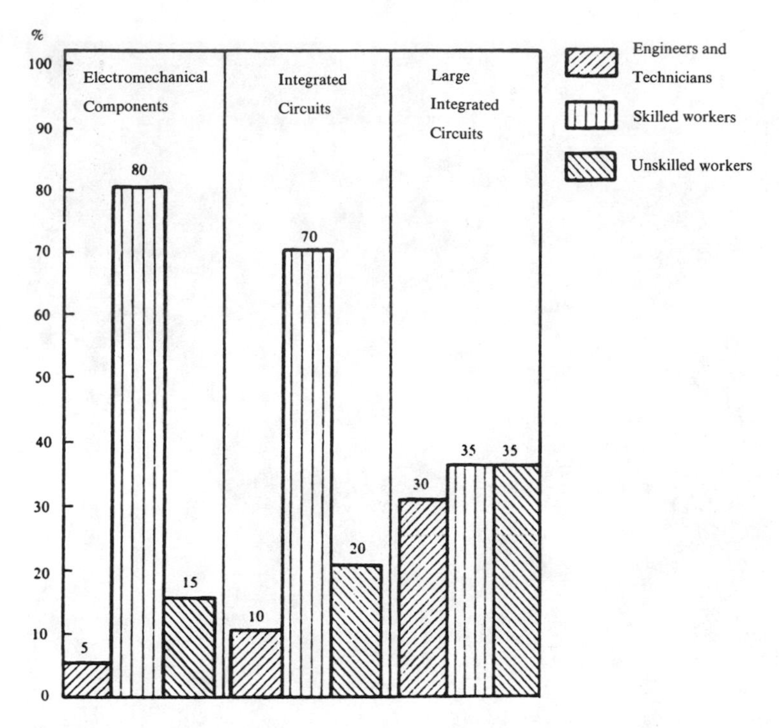

TABLE XII. Change in the employment pattern in the semiconductor industry with increasing integration of components and growing automation of their manufacturing.

There is another side, however. The number of unskilled workers employed in the electronics industry has increased. In Silicon Valley this group consists mainly of women, Mexican immigrants, Vietnamese "boat people," Filipinos, and Taiwanese. In order to place the "employment effect" of microelectronics in the best possible light, these very poorly paid jobs are counted under the heading of "high technology" just like the highly paid scientific positions (see Figs. 71 and 72). Employment in the electronics industry in the past 10 years has led to increased polarization in the American society. Many new jobs were created at the borderline of poverty, and few at the borderline of wealth. The director of the Social Welfare Research Institute in Boston noted in this connection:

FIG. 71. As progress in electronic components moved from the vacuum tube to highly integrated circuits there was a concomitant change in jobs required to manufacture the elements, an important step in the manufacturing of vacuum tubes.

We are on the way toward a low-pay society, and the new technologies are playing an important part in this. Some 67% of the jobs created since 1969 are paid at an annual level below $13 600; in the year 1969 the figure was only 45% (everything measured in 1980 dollars). The (annual income) in 1980 lay about $1000 above the officially designated poverty level" (Ref. 166, p. 206).

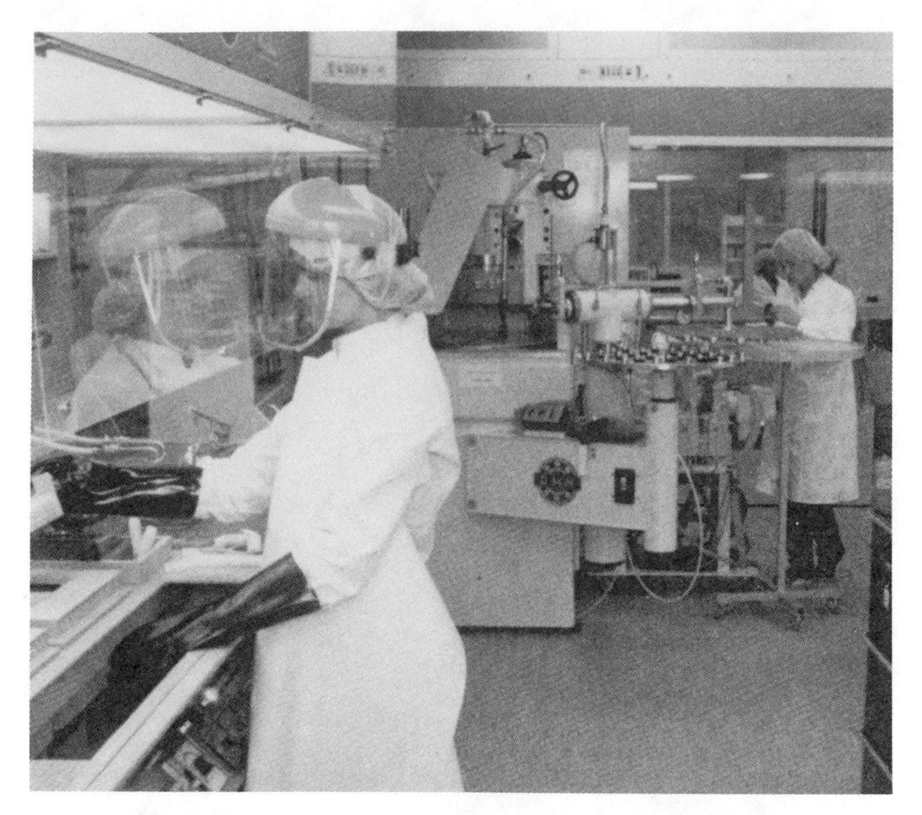

FIG. 72. The manufacturing of highly integrated silicon circuits requires strict cleanliness and absolute freedom from dust. Filtered air and protective clothing create an atmosphere which exceeds that of the operating room in sterility.

These figures did not include all unskilled workers in the field of microelectronics, because they did not take into account piece work at home, and practically all electronics firms maintained production facilities in the low-pay countries of the Pacific basin.

6

THE SCIENTIFIC ENTERPRISE: 1900 AND TODAY

THE STEREOTYPIC "IVORY TOWER OF SCIENCE," a concept fostered by romantic descriptions of reclusive researchers from Newton to Einstein, is a far cry from the everyday reality of scientific research. The story presented in this book gives an entirely different picture. At least since the existence of organized scientific work the activity of the "natural scientist," "physicist," "materials scientist," "solid-state electronics specialist," or whatever his designation, has been the focus of powerful interests, of bureaucracies and private institutions established specifically for this purpose, organizations that are pursuing concrete goals in their support of science. For the majority of career researchers their everyday activity in the "scientific enterprise" was never idyllic. A few examples and figures from the world of research in 1900 and 1970 will demonstrate once again two characteristic stages along the path of the natural scientist from the scholar's study to the industrial laboratory.

The physics profession as it was in 1900 can be shown by statistics. Schools and universities offered the main employment opportunities. According to a tabulation made in 1909 most physicists outside schools and universities were working in the traditional scientific field of astronomy, as well as in the fields of electrical engineering, meteorology, physical chemistry, and mechanics, areas of great importance as industrialization increased. Germany was the leading scientific great power. The dualism of universities and institutes of technology in Germany provided more positions than in other countries. The following are some statistical figures for the year 1900: 1078 persons were registered as professional physicists worldwide; of this number 235 were in Germany, 195 in the United States, 171 in Great Britain, and 145 in France (Ref. 45, p. 12). In 1909 the number of physicists registered had reached 3170. At this time Germany led with 962 physicists, still well ahead of Great Britain (408), the United States

(404), and France (316). Universities and upper schools still offered the most opportunities for employment (Ref. 45, p. 34ff.) Research work was performed mainly by assistants, unsalaried lecturers, and extraordinary professors, to whom fell the role of "academic proletariat" in the strictly hierarchical professors' universities. Max Weber, in his *Science as Profession* published in 1919, called the university institutes:

> enterprises of state capitalism. They cannot be managed without enormous resources. And they exhibit the same characteristics as we find in all capitalistic operations; the separation of the worker from the means of production. The worker, here the assistant, is assigned to the equipment and other means supplied by the state. He is thus as dependent on the director of the institute as an employee in a factory. (The director of the institute considers this institute to be "his" institute and he rules within it.) The assistant's position is often precarious like every "proletarian" existence... (Ref. 199, p. 7).

The daily routine of such a "proletarian" existence around the turn of the century is exemplified in the travails of Röntgen's assistant. Ludwig Zehnder, who earned his doctorate under Röntgen in 1887 and had served him, with interruptions, as assistant, unsalaried lecturer, and finally as extraordinary professor, describes several episodes which shed light on Röntgen's authoritarian and often inconsiderate behavior as well as the dependency relationship of his assistants. The account mixes personal expressions of friendship and respect with submissiveness and timid protest:

> "L. Z! Next spring to Munich!" With these words Röntgen informed his assistant on December 9, 1899, of his appointment to the University of Munich. Zehnder accompanied his master: "Soon after, when the heavy freeze had settled in, he traveled (first class!) to Munich and took me along, but to my utter amazement he immediately jarred open both windows. He was an avid hunter and was wearing his heavy fur coat, while I was wearing a lightweight ulster. But I was nine years younger and could stand the cold better. He wanted to pay my future institute a thoroughgoing visit in my company and felt that I could help him in practical matters this way."

Once when Röntgen forgot to take care of contract renewals for his assistants, he excused himself in a letter from his vacation spot in Florence: "I'm sorry that I got you into difficulty, and I would like you to let me know by writing me here as soon as possible how much you need to pay the rent... ." In 1904 there was an altercation between Zehnder and Röntgen. "First in Giessen and then in Würzburg my assistant's duties included carefully dusting off once a year, at the beginning of the holidays, the physics equipment stored in the cabinets of the institute. In Munich Röntgen had this done by the second and third assistants, while I was to supervise... ." When the latter resisted his supervision and Zehnder did not push hard enough to get the work done, he was scolded by Röntgen in such a humiliating way that he resigned his position as assistant and took a job as laboratory instructor at the Telegraph Testing Office in Berlin. In a later discussion by mail the matter was again "cleared up." Röntgen wrote: "Falling into disfavor is absolutely not the question here. Rather in my statements to you I did nothing more than my duty required. I will not retract a single syllable of what I said in my *official* capacity... ." When Zehnder then complained about his general situation as unsalaried lecturer, he had to take notice of his former master, "...The second point made in your letter, which in my opinion should be corrected, is your statement about the situation of the unsalaried lecturer and the honorary extraordinary professor at the university. You say that many people are merely 'tolerated.' I don't know what would justify saying this; the expression is somewhat reminiscent of Social Democratic slogans. The contribution of unsalaried lecturers and extraordinary professors, especially older ones, when they are good are recognized just as much as those of the other teachers at the university. The fact that they play smaller roles on the university teams is natural. The fact that they are not paid by the state is lamentable... ." This offended Zehnder more deeply than the other humiliations. He replied, "I cannot accept this reproach. My entire philosophy is contrary to the Social Democratic viewpoint. I haven't the least thing in common with Social Democracy. And except for you no one has ever accused me of using slogans! My choice of the word 'tolerated' is not unjustified, as you know. Have you perhaps forgotten that I once had to leave Würzburg in order to be able to qualify as a lecturer; that I was leaving Freiburg as extraordinary professor and reporting to take up duties as your assistant in Würzburg; that following your inquiries among the faculty I wasn't to be taken on even as unpaid lecturer without special formalities until I took over M. Wien's announced lecture following his move to Aachen, and then I was taken on? At the time we both moved to Munich you told me that there were also difficulties in getting me appointed to the Munich faculty. In Berlin meanwhile I was not allowed to lecture at the university (as a guest). More than that, I had to qualify again for this purpose (for the fifth time).... All this has caused me to feel that an unpaid lecturer is a person more tolerated than wanted. I can't believe that you would call this 'acceptance with open arms' (Ref. 207, pp. 76–102).

Seventy or eighty years later the typical physicist no longer was working in a German university or professional school but in an industrial laboratory, which in most cases belonged to a firm in the United States. In the field of electronics, for example, American industry during the 1970s conducted most of the research and development (R&D). Industry received 80% of all government R&D funding, while universities and government laboratories played a secondary role with only about 10%. As in the universities, highly paid top positions are extremely rare in the industrial laboratories, the kind of positions which helped give a region like Silicon Valley its strong drawing power for scientist, engineers, and business people and made the name of this valley a magic word (Fig. 73). Only a few scientists and engineers were able to achieve a lengendary rise like that of the two founders of the Apple Computer firm, who became multimillionaires before they were 30 years old. Most work under enormous pressure to perform; workdays of 12 and 14 h are common, as is work on weekends. Often an appointment is tied to a project task with a deadline, and after several years of intensive work, when the team is dissolved, many are burned out, a word heard of frequently in Silicon Valley, or because of overspecialization they can find no follow-on employment. Such conditions are accepted because an army of competitors is standing by to take advantage of the failure of a colleague. Best opportunities for employment are available to new universi-

FIG. 73. View of the manufacturing facilities in Silicon Valley. The speed with which some businessmen got rich gave the name of the valley a certain "magic" quality.

ty graduates and those who have gained a few years of experience in industry. It is not unusual for the average age of a research team to be under 30. Forty- and fifty-year-olds can no longer stand up to the intense stress of the marketplace. The direct competition in the research laboratories is capped by the annual "merit reviews," when management evaluates the performance of the individual researchers. These performance reviews are not only typical of the young business man in Silicon Valley; they are also conducted, for example, at Bell Laboratories:

> The first-line supervisors complete their reviews of the positions under them first. Their responsibility is to rank the research staff under their supervision... . Then everything moves to the next round and the same thing takes place one level higher, but now much larger numbers are involved. In this way every supervisor of even larger groups is provided accurate information on the people and their performance, the assignments, and the talent in other groups... . After this "merit review" and a review of earnings, the employee is assigned to a salary group (Ref. 160, p. 126).

Of course the U.S. position in physics research is not based primarily on effective utilization of scientists but is the result of a situation which developed at the end of the 1920s and was firmly entrenched by World War II. The end of World War II bestowed upon the United States absolute political and economic preeminence among world powers. The United States was the only country that did not have to convert its entire economy to war production but rather could run arms production as a business, mainly by supplying weapons to its allies. The European countries had to borrow from the United States for reconstruction after the war and the dollar became the international currency. The United States saw to it that its position was maintained in multinational bodies and agreements like the General Agreements on Tariffs and Trade, IWF, and the Organization for Economic Cooperation and Development.[72]

In order to provide their high-technology industries with the basis for success in international competition and to retain superiority in the East–West conflict, the industrial states, with the United States at the forefront, supported research and development (R&D). The United States provided more than 80% of the total expenditures for R&D by all 18 OECD states in 1967, and in 1975 it was still more than 70% (Ref. 134, p. 59). The largest expenditures in 1975 (some 55% of all American R&D money) were allotted to the area of aviation and aircraft, which was an important area for electronics as well. (The trend, however, was downward.) The next largest amount of funding went to electronic components and telecommunications equipment, with more than 30% of the total funding, and the trend in this case was upward (Ref. 134, p. 50). Although overall funding of R&D in all areas of production was approximately evenly divided between government and industry, in the two fields mentioned above the public supported the bulk of the effort (up to 80%). The largest source of funds was the Defense Department with 70% of all government-supported R&D in industry, and NASA disbursed an additional 18% (Ref. 3, p. 275). In most cases the industrial contractor was guaranteed a profit as a result of its research effect and the government assumed the risk of cost over-runs. In addition, the knowledge and experience gained during the R&D process guaranteed the performing firm a competitive advantage.[32] In this way the costs of technical innovations in the entire field of electronics (military and civilian) were in a large measure borne by government organizations and thus shifted to all taxpayers.

In recent years the U.S. government has reduced the civilian R&D budget while sharply increasing outlays for military research (Table XIII). The proportion of scientists working on military research worldwide, currently estimated at about 50%, is thus continuing to rise. President Reagan's 1983 announcement of his Strategic Defense Initiative (SDI) indicates that this growth will continue in the next decade. The objective is to develop laser or particle weapons that will make the Russian intercontinental missiles largely ineffective, so that the weapons race between East and West will ultimately be decided in favor of the West. Since the planned system will be stationed in space and the space

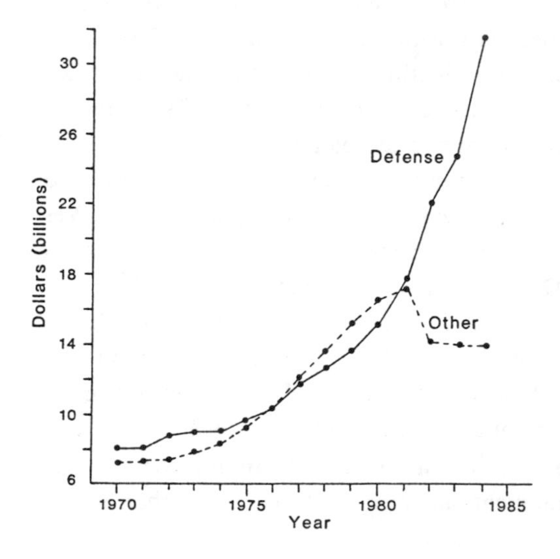

TABLE XIII. The U.S. research and development budget broken down into military ("Defense") and civilian ("Other") components (Ref. 172).

research sector is closely associated with the development of the new technologies, companies in this field are most likely to be awarded lucrative research contracts.

The radiation weapons used in the present arms buildup would not be conceivable without the practical experience and equipment of elementary particle physics, which at the same time is considered a prime example of "pure" basic research not directed toward achieving some practical objective. The sacred search for knowledge and truth can be motive enough for the research people in their work, but the situation is different with the people in government and industry who let the contracts. An important factor in the government's attitude is the experience that military technology, especially since the Second World War, is increasingly dependent on the vitality of the basic sciences. Recognizing this, at its summit conference in 1957 NATO established a Science Committee which is supposed to support and coordinate basic research in the NATO countries. Another government stake in basic research is the strength of the national economy in international competition. Here government and industry see a common basis for supporting "pure" science:

> The first and most immediate basis lies in the fact that basic information is useful. Most industries involved in rapid technological change utilize technologies arising from important areas of science, as those who make chemical or electronic products. In these and other branches of industry both the government and industry vigorously and sucessfully support the sciences which are of importance for them.
>
> New studies show...the important role played by scientific knowledge as a sort of infrastructure for the work of engineers and inventors and the influence it has on the course of that work. The studies also show that some of the basic innovations of the 20th century, which opened new product markets, were actually stimulated by scientific advances. Finally, we know from historical experience that industrial practice frequently requires theoretical explanation and thus challenges basic research... .
>
> The second basis for government support of basic research is the fact that the market mechanism would lead to inadequate investment in research. There are several reasons for this; industry doesn't look as far ahead, and it would be impossible for private investors to utilize themselves of all the results of their own basic research. The market mechanism is not capable of assuring a socially desirable level of basic research (Ref. 134, p. 62ff).

The example of CERN (Centre Européen pour la Recherche Nucleaire), the large European research facility for elementary particle physics, can illustrate where the motives for such basic research lie. CERN is usually associated with terms like "quark," "charm," "quantum chromodynamics," and other exotic concepts from an area of physics which has become completely unintelligible to the layman, and not with the applications-oriented physics which is the subject of this book.

The fact that this branch of physics as an area of advanced technology also plays a critical role in the applications area has been shown by a study on the "economic benefit" of CERN:

> CERN was established about 30 years ago by several European countries which wanted to make it possible, through this community project, for Europe to regain a leading position in particle physics. In pursuing this objective CERN since that time has built a series of advanced accelerators and research devices. This required the investment of significant resources, as a result of which CERN, in addition to its significant contributions to science, has also assumed an important function in the technological and economic areas as well... . As a rule a large portion of the scientific equipment required for its research has been supplied by industry. Since the specifications and requirements frequently exceed the "knowhow" available, they represent a challenge for the manufacturer. The positive results arising from this situation, such as new products, quality improvements, higher productivity, etc., can be called "secondary" economic effects... .
>
> The present report describes the methods and substantive results of a study to quantify secondary economic benefits. These benefits are defined as the total increase in business turnover and cost savings which have taken place in the European advanced technology industry as a result of CERN contracts between 1973 and 1982... .
>
> Tabulation of the benefits reported by the surveyed firms for all 519 technology suppliers gives a total benefit for the 1973–1982 period of 4.8 billion Swiss francs (at 1982 prices)... . One Swiss franc expended by CERN for advanced technology created an average of three Swiss francs in economic benefit (Ref. 24, p. VIff).

It is important to keep in mind that the technological knowhow discussed in this report is of interest to the government agencies funding such basic research even if initially the actual leverage and economic advantage in international competition cannot yet be defined. Tables XIV and XV show which branches of industry generate "economic benefits" from the tax dollars so generously channeled by the government into this research. The gross differences between sciences as it was practiced in 1900 and the situation today should not lead a person to the misconception that anything substantial has changed in the setting of goals for science. If disciplines like electronics and low-temperature physics developed from a stage of tinkering theoretically and experimentally to almost universally useful advanced technologies, this did not happen as a result of growth following an independent and purely internal logic separate from

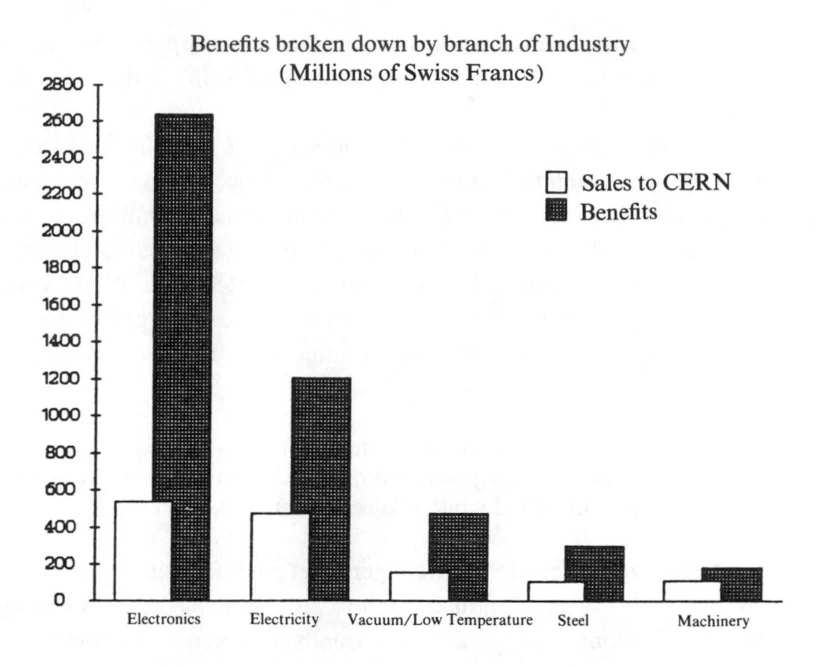

TABLE XIV. Total sales and total profits of 519 high-technology CERN suppliers, broken down by branch of industry.

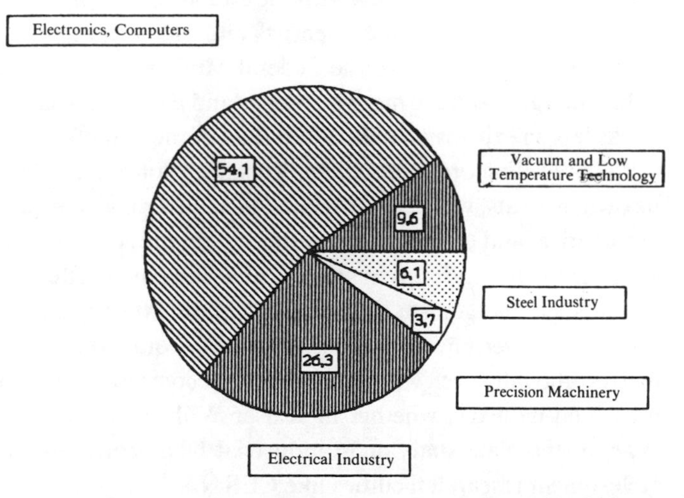

TABLE XV. Percentage breakdown of profits earned by 519 high-technology vendors, by branch of industry.

external events, but on the basis of planned progress supported by government and industry, which are making more use of scientific findings for their own purposes.

The popular scientific literature and speeches by politicians still give the public the impression that at least the science concerned with basic research is "free" and guided only the search for truth and the latest findings. Things sound much different when research planners talk among themselves. According to a member of the Research and Technology Committee of the West German Parliament, the prevailing view is that in basic research "government control should be as small as possible"; at the same time, however, the same science policy maker states:

> Basic research increases further in the research budget... . We see a gradation in opportunities for government control and influence from basic research to very directly defined projects (Ref. 187, p. 118).

And scientists themselves are eager to set goals for science in close cooperation with politicians and industry and to lay out steps on the way so that the political and economic objectives are usually accepted as the standard and are sanctioned with scientific authority. The larger and more useful science becomes, the heavier its involvement in industrial management and politics. Thus, in 1957 in the United States, as a result of "Sputnik shock," an elite group of scientists (the President's Scientific Advisory Committee) was given direct access to the White House. In West Germany scientific advice to the government was institutionalized in 1956 with the establishment of the German Atom Commission, a panel of leading scientists and industrialists in the Atomic Energy Ministry, the precursor of the Federal Ministry of Research and Technology. While this gave science more resources and more of a voice in decision-making, it was less an expression of freedom than of the voluntary adjustment of science to the prevailing order in society. This does not prevent science from pursuing its own interests, which in individual instances may even run counter to prevailing political and economic interests. In this the scientific community is no different from other groups in society. One can hardly cite instances, however, in which such antagonism of interests has disturbed the broad conformance of science to government interests. Quite the contrary. The physics community has traditionally been well known for its accommodation and loyalty to government and industry, whether in Kaiser Wilhelm's empire, in the First World War, in the Nazi state, or in industrial laboratories like Bell Labs or major government research facilities like CERN.

APPENDIX

ELECTRONS IN THE CRYSTAL LATTICE

An electron in the isolated atom is allowed only those "states" in which it possesses specific energy values derived from quantum mechanics [Fig. 74(a)]. When several atoms are joined together to make a molecule, the number of energy states allowed for an electron in the molecule increases; the energy levels of the individual atoms are split by the overlapping of the atomic fields [Fig. 74(b)]. In a crystal, the overlapping of neighboring atomic fields leads finally to a splitting into "bands" of energy states, which are separated from one another by zones of "forbidden" energy regions [Fig. 74(c)].

The distribution of electrons among the energy states in a "band" follows the Pauli principle, the "housing office" for electrons (according to Pauli and Sommerfeld): each quantum state may be occupied by only one electron. If we imagine the energy states of a crystal to be filled one after the other by the available electrons, then the "last" electron will assume an energy value which has only empty states "above" it and fully occupied states below it in the series.

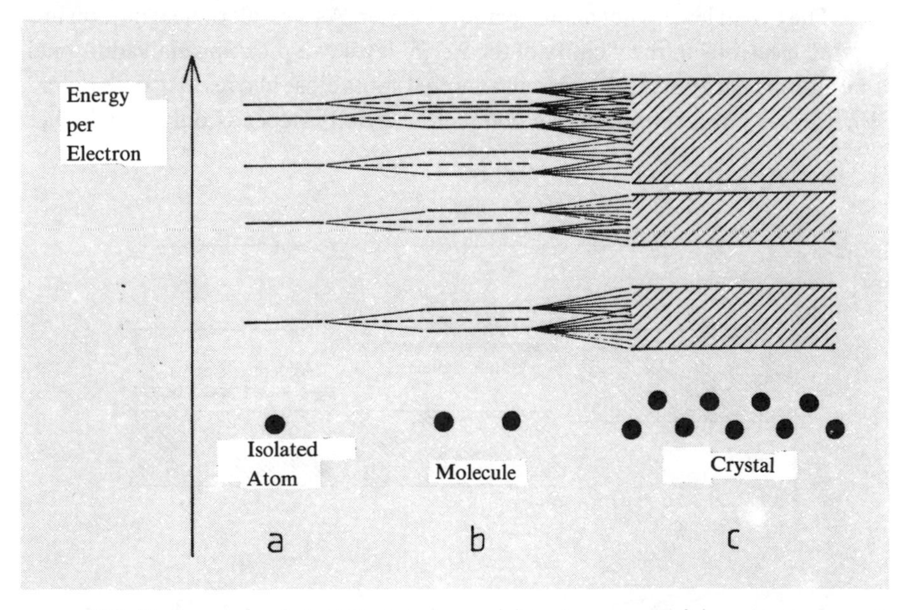

FIG. 74. Energy levels in the isolated atom (a), in the molecule (b), and in the crystal (c).

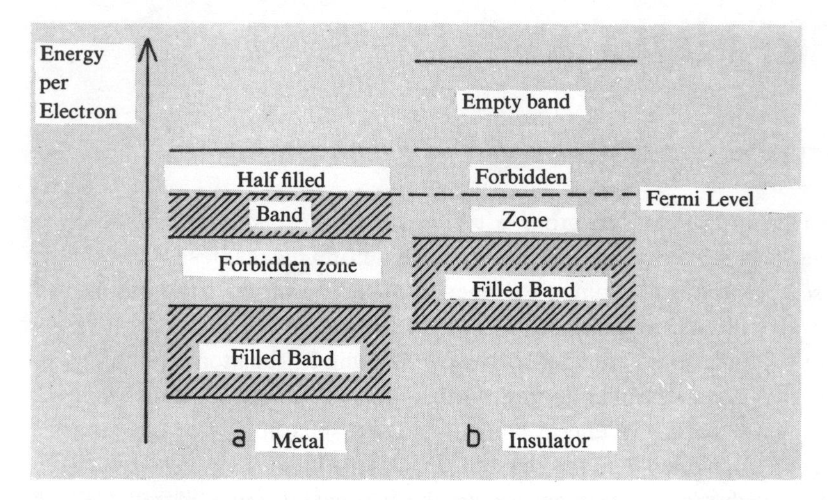

FIG. 75. Energy bands in a metal (a) and in an insulator (b).

This limiting value is called the Fermi level. Depending on the composition of the crystal this energy level is located in a band of permitted or forbidden states. In the former case we are dealing with a metal [Fig. 75(a)], and in the second with an insulator or a semiconductor [Fig. 75(b)].

The Fermi level is not an insuperable barrier. As a result of heat motion in a crystal, electrons in the vicinity of the Fermi level can pick up small additional amounts of energy and take over unoccupied spots close above this "threshold." Whether or not a substance can conduct electricity depends only on whether

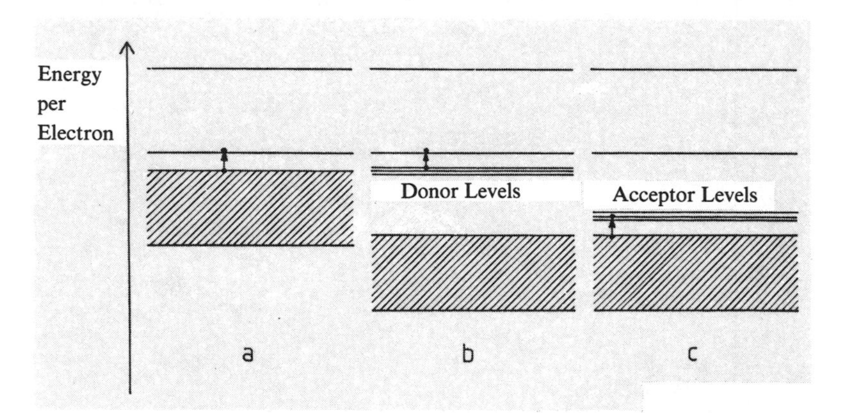

FIG. 76. Energy bands in a semiconductor; (a) intrinsic semiconductor, (b) n conductor, (c) p conductor.

electrons can change their places within bands (as in a metal) or between bands under the influence of an external electrical voltage. Electron movement between bands is a characteristic of semiconductors. The following three possibilities can be described: (1) The energy gap between the last occupied band and the following empty band is small enough to be leaped by the electrons from the filled band when they take up heat energy [Fig. 76(a)]. This type is called an "intrinsic semiconductor." (2) Additional energy levels are created within a forbidden zone by the foreign atoms ("doping"), either directly below the first unoccupied band, so that electrons can make their way from that point to the empty band [Fig. 76(b)], or so that electrons can be removed from it leaving "holes" in the otherwise thickly populated band [Fig. 76(c)]. In case (b) we are talking about negative conduction (*n* conduction), while in case (c) we have positive conduction (*p* connection), since here the electrical current flows via missing electrons or "holes," which is equivalent to an effective transport of positive charges.

Figures 74–76 are merely schematic representations of the electron activity in a crystal. Actually, the zones of allowed and forbidden energy ranges are three-dimensional constructs dependent on the ("Brillouin zones"). The same thing is true for the Fermi level, which actually describes the surface of a three-dimensional volume that sometimes will have a bizarre appearance because of the particular lattice structure involved (the "Fermi surface").

The examples of Brillouin zones and Fermi surfaces shown in Figs. 77 and 78 illustrate the actual relationships in crystals. It should be noted that here we

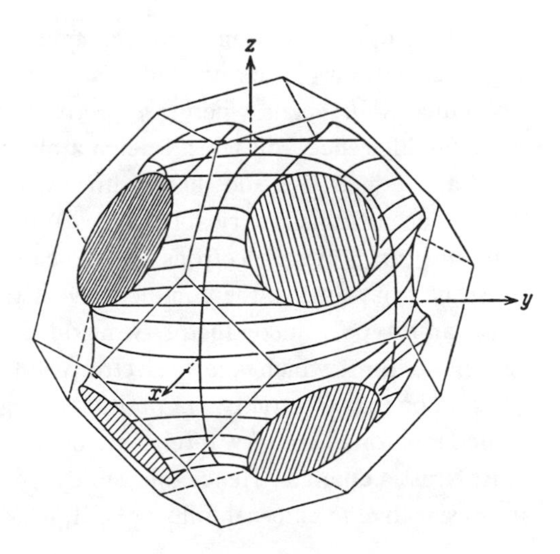

FIG. 77. Brillouin zone with Fermi surface.

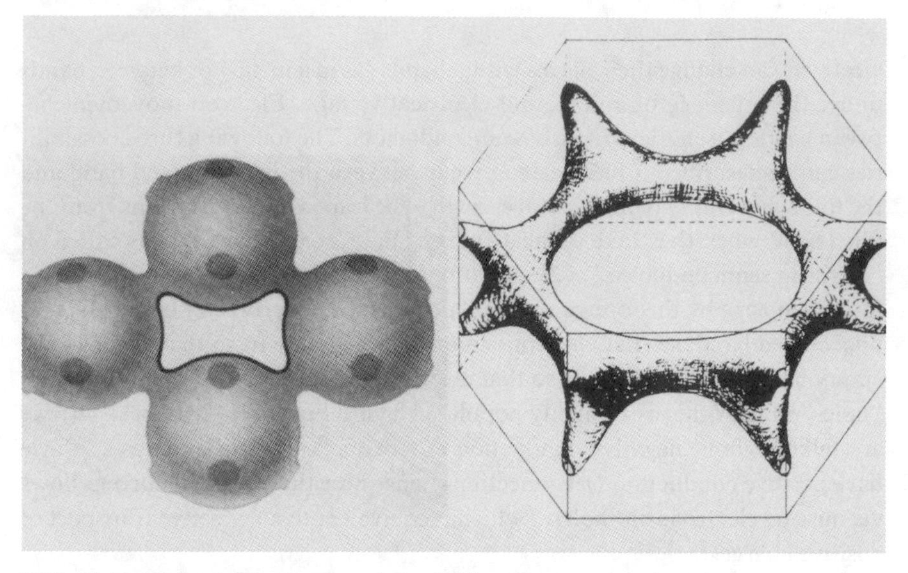

FIG. 78. Fermi surfaces of gold or copper (left) and magnesium (right).

are not dealing with "visible" constructs in our ordinary spatial environment but with structures in "wave-number space." This is not the ordinary space spanned up by the coordinates of the sites of the electrons but by their velocities. Further details and explanations must be left to textbooks.

THE PRINCIPLE OF THE TRANSISTOR

The properties of semiconductor materials depend heavily on impurities and deformations of their crystalline structure, quite in contrast to the properties of metals. In metals, where free electrons are so abundant, a few more or less make no difference. But when foreign atoms that can give up or take up electrons are lodged in a substance with few charge carriers of its own, major changes occur in its electrical behavior. When the foreign atoms in the crystal structure can give up electrons (the so-called donor atoms), additional free electrons can increase the conductivity of the substance and the material is called an *n*-type semiconductor. Semiconductors containing acceptor impurities, that is, those which take up electrons and cause a deficit of electrons (called holes), conduct electricity via positive charge carriers and are called *p*-type semiconductors. Just a few foreign atoms per million semiconductor atoms can cause drastic changes. Hence one can easily understand why semiconductors are so sensitive to external influences. If, for example, light falls on a crystal, a

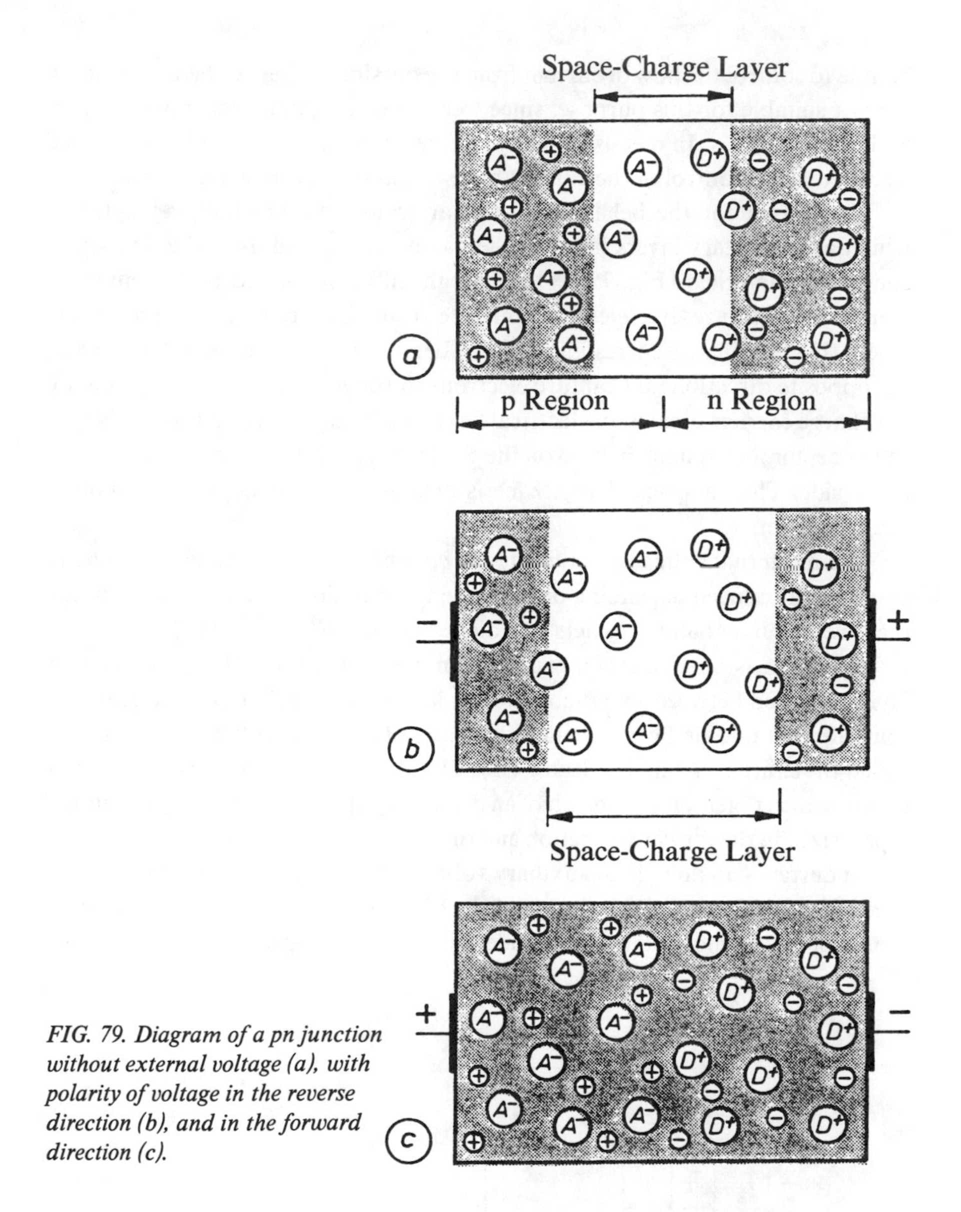

FIG. 79. Diagram of a pn junction without external voltage (a), with polarity of voltage in the reverse direction (b), and in the forward direction (c).

light quantum can transfer its energy to a tightly bound electron and free it from its bond. It is then available for current conduction and the conductivity increases.

A material whose conductivity can be altered easily appeared ideal for use in new kinds of switches and even amplifiers. Researchers felt that they should be able to control the flow of current from the outside as they wished. Insulators are not suitable for this purpose, since free electrons can be generated only at high temperatures. In metals so many charge carriers are available to conduct electricity that control of the current by external means appeared hopeless.

Important for the behavior of bipolar transistors are the *pn* junctions, which are boundary layers between the *p*-conducting and *n*-conducting semiconductor materials [Fig. 79(a)]. Through diffusion caused by the environmental temperature, free electrons migrate from the *n* region to the *p* region, where they eliminate a corresponding number of holes, while free holes move in the opposite direction and capture electrons in the *n* region. The movement of the charge carriers creates an electrical field which opposes the diffusion. Negative acceptor ions remain behind on the *p* side and positively charged donors on the *n* side. Thus a space-charge zone is created which is depleted of mobile charge carriers.

If an external voltage is applied to the *pn* junction, the space-charge layer is enlarged or reduced depending on polarity; the junction obstructs or facilitates the flow of current and thus acts as a diode [Figs. 79(b) and 79(c)].

In a transistor such as, for example, an *npn* transistor, a thin *p*-conductive layer is lodged between two *n*-conducting layers (Fig. 80). For historical reasons the thin middle layer is called the base (*B*) and the other two layers are called the emitter (*E*) and collector (*C*). The *npn* transistor is equivalent to two diodes joined together. As a result of an externally applied voltage (*u*) one diode is polarized in the forward direction and the other in the closed direction, so that no net current can flow. If an auxiliary voltage (*u*) is applied to the lower diode (see Fig. 72) between the emitter and the base in the forward direction, a current of electrons moves into the boundary layer of the upper diode, which is still

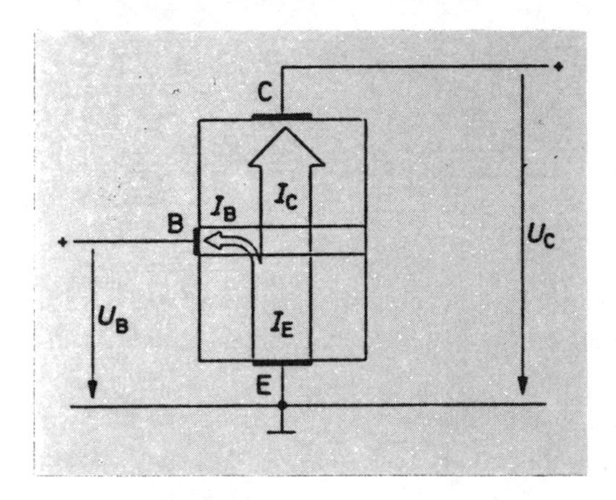

FIG. 80. Current flow in an npn transistor.

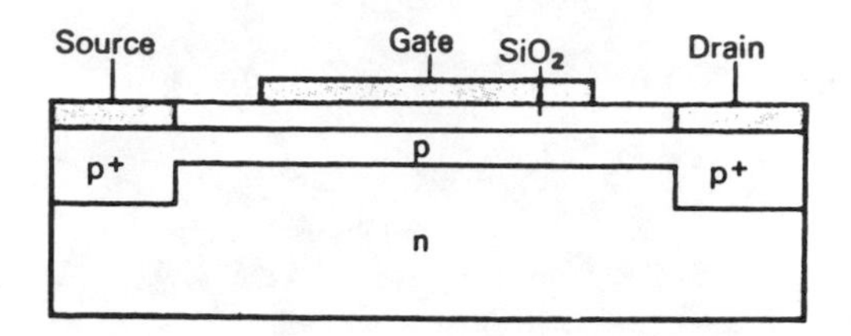

FIG. 81. Diagram of a field-effect transistor.

polarized in the nonconducting direction. The barrier layer, however, acts only against electrons flowing from the collector in the direction of the base, and not against the electrons important here which cross the *pn* junction in the base-collector direction, so that the latter can flow off into the collector current circuit. Applying the auxiliary voltage, therefore, causes current to flow through the transistor. Even small changes in voltage between the emitter and the base change the number of electrons that pass the *pn* junction and cause a major change in the flow of current through the transistor. Thus a current flows through the diode connected in the reverse direction, because the forward-operating diode is located in the immediate vicinity. The distance amounts only to a thousandth of a millimeter.

The mode of operation of the field-effect transistor is much simpler. Like the bipolar transistor, this one has three electrodes, called the source, the gate, and the drain. As an example we show a *p*-channel field-effect transistor (Fig. 81). A long narrow *p* zone (channel) is diffused into an *n*-conducting silicon crystal, both ends of which terminate in heavily doped *p*-conductive layers that are provided with source and drain electrodes. The third electrode, the gate, covers the channel, but is separated by an insulating layer of oxidized silicon (SiO_2). The gate and the channel together constitute a condenser with the insulating layer as the dielectric. If the gate is negatively charged, positive charge carriers collect in the *p*-conducting channel, increasing the conductivity. Varying the voltage controls the conductivity and thus the current flow between source and drain. Heavy current flows can be controlled with small voltages.

REFERENCES

Sources used:

AHQP = Archive for the History of Quantum Physics. A copy of this microfilm collection is in the Deutsches Museum.
SAA = Siemens-Akten-Archiv (Siemens Document Archive), Munich.
SSP = Collection of source materials for the project "History of Solid State Physics," Deutsches Museum, Munich.

1. Achilles, M. and R. W. Pohl, 1977, Praxis der Naturwissenschaften, **26**, 155–159.
2. Alexander, W. O., 1978. *The utilization of metals.* In: T. I. Williams (Ed.), A History of Technology. Vol. VIII, 1900–1950, Part I, p. 427–461. (Oxford Univ. Press, N.Y.).
3. Atkins, A. G. (Ed.). 1975. Guide to world science. *Vol. 22: United States of America.* Guernsey, Channel Islands.
4. Böhme, D. (Ed.). 1975. Proclamations and speeches of German professors in the First World War (*Aufrufe und Reden deutscher Professoren im Ersten Weltkrieg*). Stuttgart.
5. Bacon, R. F., C. E. K. Mees, W. H. Walker, M. C. Whitaker, and W. R. Whitney. 1917. Research in industrial laboratories. Science: 34–39.
6. Bardeen, J. 1977. Interview by L. Hoddeson. SSP.
7. Bell Laboratories Record, 1952. Vol. 30.
8. Bell Telephone Quarterly, 1929. Vol. 8.
9. Benz, U. 1973. Arnold Sommerfeld: A Scientific Biography. Dissertation, University of Stuttgart.
10. Bernal, J. D. 1978. Social history of the sciences (*Sozialgeschichte der Wissenschaften*). New edition of the pocket edition which appeared in 1970 under the title *Science* (*Wissenschaft*). Rowohlt TB/6224–6227. Reinbek.
11. Beyerchen, A. D. 1982. Scientists under Hitler; Physicists in the Third Reich (*Wissenschaftler unter Hitler; Physiker im Dritten Reich*). Ullstein TB 34018. Frankfurt Am-Main. (Original American edition 1977.)
12. Birr, K. 1957. *Pioneering in Industrial Research.* Washington.
13. Bley, C. 1949. Radar Secret (*Geheimnis Radar*). Hamburg.
14. Brachner, A., G. Hartl, and S. Hladky. 1983. Atomic, Nuclear, and Elementary particle Physics (*Atom-,Kern-, Elementarteilchen Physik*). Deutsches Museum, Munich.
15. Bragg, W. Bragg papers, Royal Institution, London.
16. Brandt, L. 1967. On the history of radar technology in Germany and Great Britain (Zur Geschichte der Radartechnik in Deutschland und Grossbritannien). Paper presented at the XV Convegno Internazionale delle Comunicazioni, Genova.
17. Braun, E., and S. MacDonald. 1982. *Revolution in miniature.* Cambridge University Press, N.Y.).

18. Brecht, B. 1967. (The Life of Galileo. In: Bertold Brecht, Collected Works, Vol. 3), (*Bertold Brecht, Gesammelte Werke, Bd. 3*). Suhrkamp Werkausgabe, Frankfurt Am-Main.
19. Burchardt, L. 1975. (Science policy in Kaiser Wilhelm's Germany; prehistory, establishment, and growth of the Kaiser-Wilhelm Society for the Advancement of Science. In: W. Treue (Ed.), *Studies on Natural Science, Technology, and Economics in the 19th Century*, Vol. 1. (Studien zu Naturwissenschaft, Technik, and Wirtschaft im 19. Jahrhundert, Band I). Göttingen.
20. Busch, A. 1959. The history of the unsalaried university lecturer; a sociological study on the evolution of the German university as a large-scale enterprise (*Die Geschichte des Privatdozenten; eine sociologische Studie zur grossbetrieblichen Entwicklung der deutschen Universitäten*). Stuttgart.
21. Cahan, D. L. 1980. The Physikalisch-Technische Reichsanstalt: A Study in the Relations of Science, Technology, and Industry in Imperial Germany. Dissertation, Johns Hopkins University, Baltimore.
22. Cardwell, D. S. L. 1982. *The organization of science in England.* London.
23. Carnot, S. 1824. Reflections on the motive power of fire and machines suitable to develop this power. "(Réflexions sur la puissance motrice du feu et sur les machines propres à développer cette puissance)." Paris. (Quotation from German translation by W.Ostwald, Leipzig, 1909.)
24. CERN. 1985. The economic benefit of CERN contracts to industry, second study (Der wirtschaftliche Nutzen der CERN-Aufträge an die Industrie, zweite Studie). CERN Report 85-04. Geneva.
25. ———. 1981. Once upon a time...The history of the microprocessor. Chip, **11**, 16–21.
26. Czada, P. 1969. The Berlin electrical industry during the Weimar period, a regional statistical and economic history study. In: *Einzelveröffentlichungen der Historischen Kommission zu Berlin*, No. 4.
27. Debye, P. 1965. Interview by D. M. Kerr and L. P. Williams, 22 December 1965. (With kind permission of Donald M. Kerr.)
28. Documents concerning the establishment of the Kaiser Wilhelm Society and the Max–Planck Society for the Advancement of Science (Dokumente zur Gründung der Kaiser-Wilhelm Gesellschaft und der Max–Planck Gesellschaft zur Förderung der Wissenschaften). An exhibition catalog.
29. Duisberg, C. 1920–21. Duisberg to Wien, 1920–1921. Bequest of Willy Wien in the Deutsches Museum.
30. Eckert, M., W. Pricha, H. Schubert, and G. Torkar. 1984. Confidential Advisor Sommerfeld—theoretical physicist; documents from his papers (Geheimrat Sommerfeld—Theoretischer Physiker, Eine Dokumentation aus seinem Nachlass). Munich.
31. École Polytechnique. 1897. One-Hundredth Anniversary of the Ecole Polytechnique, 1794–1894 (*École Polytechnique: Livre du Centennaire 1794–1894*). 3 Vols. Paris.

32. European Community. 1969. EC-Study: Research and development in the field of electronics in the European Community countries and in the most important outside nations (*EG-Studie: Die Forschung und Entwicklung auf dem Gebiete der Elektronik in den Ländern der Gemeinschaft und in den wichtigsten Drittländern*). Brussels.
33. Einstein, A. 1907. The Planck theory of radiation and the theory of specific heat. Annalen der Physik, **22**, 180–190.
34. Elsasser, W. M. 1978. *Memoirs of a physicist in the atomic age.* New York.
35. Esau, A. 1940. Electric waves in the centimeter region. Schriften der deutschen Akademie der Luftfahrt-forschung, 1940, No. 20:3–14.
36. Ewald, P. P. (Ed.) 1962. *Fifty years of x-ray diffraction.* Utrecht.
37. Fagan, M. D. *A history of engineering and science in the Bell Systems; national service in war and peace (1925–1975).* Bell Telephone Laboratories, Inc.
38. Ferencz, B. B. 1981. Wages of horror; Compensation refused for Jewish forced labor; a chapter of German post-war history (*Lohn des Grauens; Die verweigerte Entschädigung für jüdische Zwangarbeiter; Ein Kapitel deutscher Nachkriegsgeschichte*). Frankfurt, New York.
39. Fishcr, L. 1913. L. Fischcr to F. A. Spicckcr, 26 July 1913, SAA 68/Li 185.
40. Flechtner, H. J. 1959. Carl Duisberg; from chemist to leader of the economy (*Carl Duisberg. Vom Chemiker zum Wirtschaftsführer*). Düsseldorf.
41. Forman, P. 1967. The Environment and Practice of atomic physics in Weimar Germany. Dissertation, University of California, Berkeley.
42. Forman, P. 1971. Weimar Culture, Causality, and Quantum Theory. 1918–1927; Adaptation by German Physicists and Mathematicians to a Hostile Environment. Historical Studies in the Physical Sciences. **2**, 1–116.
43. Forman, P. 1973. *Scientific Internationalism and the Weimar Physicists: The Ideology and its Manipulation in Germany after World War I,* ISIS, 64 150–180.
44. Forman, P. 1974. The financial support and political alignment of physicists in Weimar Germany. Minerva **12**, 39–66.
45. Forman, P., J. L. Heilbron, and S. Weart. 1975. Physics ca. 1900. In: Historical Studies in the Physical Sciences, Vol. 5.
46. French, S. J. 1938. The science of alloy building. Scientific American, **158**, 78–80 and 152–155.
47. Friedrich, W. 1949. Recollections about the discovery of interference phenomena with x-rays. Naturwissenschaften, 1949: 354–356.
48. Friedrichs, G., and A. Schaff. 1984. For better or for worse: microelectronics and society (*Auf Gedeih und Verderb, Microelektronik und Gesellschaft*). Rowohlt. TB 8106, Reinbek.
49. Fröhlich, H. 1936. Electron theory of metals (*Elektronentheorie der Metalle*). Berlin.
50. Fröhlich, H. 1958. *Theory of dielectrics.* Oxford, 1940. (Second edition, 1958.)
51. Fry, A. (Ed.) 1934. The research institutions of the Krupp firm (*Die Forschungsanstalten der Firma Krupp*) Essen.
52. Funk Technik, 1953, **13**, 410.
53. Geiger, H. and K. Scheel. (Eds.) 1933. Handbook of physics (*Handbuch der Physik*). Vol. 24, Part 2: Structure of coherent matter (*Aufbau der zusammenhängenden Materie*). Edited by A. Smekal, Berlin.

54. Gelhoff. 1930. Gelhoff to Sommerfeld, December 3, 1930. Sommerfeld bequest. Deutsches Museum.
55. Gerdien, H. 1914. Gerdien, H. to Siemens, W. von, May 13 and June 4, 1914. SAA 4/LK 166 (Wilhelm v. Siemens).
56. Gerdien, H. 1918. Memorandum concerning the mission of the Physical-Chemical Laboratory. SAA 68/Li 185.
57. Gerdien, H. 1944. History of the Research Laboratory (Geschichte des Forschungslaboratorium). Unpublished manuscript. SAA 68/Li 185.
58. Gerlach, W. 1978. Robert Wichard Pohl, Jarbuch der Bayer. Akad. Wiss., **1978**, 1–6.
59. Gersdorff, K. von, and K. Grasmann. Airplane engines and jet power plants (*Flugmotoren und Strahltriebwerke*). Munich.
60. Glaser, G. 1982. Glaser, G. to Teichmann, J., May 24, 1982. SSP.
61. Glasser, O. 1931. Wilhelm Conrad Röntgen and the History of x-rays (*Wilhelm Conrad Röntgen und die Geschichte der Röntgenstrahlen*). Berlin.
62. Goetzeler, H. 1972. History of semiconductor components in electronics. Technikgeschichte, **39**, 31–50.
63. Goudsmit, S. 1947. ALSOS. New York.
64. Groves, L. 1962. Now it can be told! (Neue amerikan. Taschenbuchauflage 1983.)
65. Gudden, B. and R. Pohl. 1923. The quantum equivalent in the photoelectric conductor. Zeitschrift für Physik, **17**, 331–346.
66. Gummert, H. 1976. The development of new technical methods using scientific concepts in the area of German heavy industry, as exemplified in the case of the Krupp firm in Essen. In: Studies on natural science, technology, and economics in the 19th Century (Studien zur Naturwissenschaft, Technik und Wirtschaft im 19. Jh. Göttingen. Pages 351 ff.
67. Hallgarten, G. W. F., and J. Radkau. German Industry and Politics: From Bismarck to the Present (*Deutsche Industrie and Politik. Von Bismarck bis in die Gegenwart*). Rowohlt TB 7450. Reinbek.
68. Hanson, D. 1984. The history of microelectronics (*Die Geschichte der Mikroelektronik*). Heyne TB 15/3. Munich.
69. Hanslian, R. 1951. From gas on the battle field to atomic war (*Vom Gaskampf zum Atomkrieg*). Stuttgart.
70. Heisenberg, W. 1968. Reverberations from Sommerfeld's work into the present, Physikalische Blätter. **24**, 530–537.
71. Heisenberg, W. 1976. The part and the whole (*Der Teil und das Ganze*). Itv, TB 903. Munich. (First edition, 1969; the marginal notes in the text refer to this edition.)
72. Held, K. and T. Ebel. 1983. (War and peace (*Krieg und Frieden*) Suhrkamp TB, NF 149. Frankfurt.
73. Hellwege, K.-H. 1970. Introduction to atomic physics (*Einführung in die Atomphysik*). Heidelberg.
74. Helmholtz, H. 1894. Preface (Vorrede). Abhandlungen der Physikalisch-Technischen Reichsanstalt, **1**, II.
75. Henriksen, P. W. The emergence of solid state physics research at Purdue University during World War II. Manuscript slated for publication in: Historical Studies in the Physical Sciences.

76. Hermann, A. 1968. Albert Einstein/Arnold Sommerfeld: Correspondence; Sixty letters from the golden age of physics (*Albert Einstein/Arnold Sommerfeld: Briefwechsel. Sechzig Briefe aus dem goldenen Zeitalter der Physik*). Basel/ Stuttgart.
77. Hermann, A. 1969. Early history of the quantum theory (*Frühgeschichte der Quantentheorie*), Mosbach.
78. Hermann, A. 1975. Science helps "increase business" (Wissenschaft hilft "Gewerbe erheben"). Wirtschaft und Wissenschaft Vol. 2: 3–7.
79. Hermann, A. 1977. Science of the century: Werner Heisenberg and the physics of his time (*Die Jahrhundertwissenschaft: Werner Heisenberg und die Physik seiner Zeit*), Stuttgart.
80. Hermann, A. 1982. How science lost its innocence (*Wie Wissenschaft ihre Unschuld verlor*), Stuttgart.
81. Hermann, A., V. Weisskopf, and K. von Meyenn, 1979. Wolfgang Pauli: Scientific correspondence (*Wolfgang Pauli: Wissenschaftlicher Briefwechsel*) Vol. I, 1919–1929 (Band I, 1919–1929), New York.
82. Heywang, W. 1984. The role of physics in modern electronics. Physikalische Blätter, 40, 265–270.
83. Hilsch, R. 1934. R. Hilsch to R. W. Pohl, August 29, 1934, SSP.
84. Hilsch, R. 1934. R. Hilsch to R. W. Pohl, September 7, 1934, SSP.
85. Hilsch, R. 1939. Electron conduction in crystals. Die Naturwissenschaften, 29, 489–492.
86. Hilsch, R. 1982. Speech on the occasion of the 1965 Color Center Symposium held in Urbana, IL. Transcribed by K. Szymborsky in Vienna.
87. Hoch, P. K. 1983. The reception of central European refugee physicists. U.S.S.R., UK, U.S.A. Annals of Sciences, **401**, 217–246.
88. Hoddeson, L. 1978. Multidisciplinary research in mission oriented laboratories: the evolution of Bell Laboratories' program in basic solid-state physics, culminating in the discovery of the transistor, 1935–1948. Urbana, Illinois.
89. Hoddeson, L. 1980. The entry of the quantum theory of solids into the Bell Laboratories, 1925–1940. Minerva, **18**, 421–447.
90. Hoddeson, L. 1981a. The discovery of the point-contact transistor. Historical Studies in the Physical Sciences, **12**, 40–76.
91. Hoddeson, L. 1981b. The emergence of basic research in the Bell Telephone System, 1875–1915. Technology and Culture, **22**, 512–544.
92. Hoddeson, L., E. Braun, J. Teichmann and S. Weart 1986. The history of solid state physics (Working title). Publication slated for 1986. The book is a product of the international project entitled "History of Solid State Physics."
93. Holden, H. 1976. Interview by L. Hoddeson. SSP.
94. Hund, F. 1969. High points of physics in Göttingen. Phys. Blätter, **25**, 145–153 and 210–215.
95. Hunsicker, H. Y. and H. C. Stumpf. 1965. History of precipitation hardening. In: *The Sorby Centennial Symposium on the History of Metallurgy* (Edited by C. S. Smith). New York. pp. 271–311.
96. Irving, D. 1967. The dream of the German atomic bomb. (*Der Traum der deutschen Atombombe*). Gutersloh.

97. Ittner, W. B. and C. J. Kraus. 1961. Superconducting computers. Scientific American, July 1961, 124–136.
98. *Jahrbuch der deutschen Luftfahrtforschung* (Yearbook of German aeronautical research). 1937.
99. Jewett, F. 1932. Utilizing the results of fundamental research in the communication field. Bell Telephone Quarterly, **11**, 143–161.
100. Joffe, A. 1967. Meetings with physicists (*Begegnungen mit Physikern*). Basel.
101. Johnson, V. A. 1969. *Karl Lark-Horovitz, Pioneer in solid state physics*, Oxford.
102. Jones, H. 1980. Notes on work at the University of Bristol. Proceedings of the Royal Society of London, A **371**, 52–55.
103. Jungk, R. 1956. Brighter than a thousand suns, the destiny of the atomic researchers (*Heller als tausend Sonnen. Das Schicksal der Atomforscher*). Stuttgart and Bern.
104. Kaempffert, W. 1911. Made in Germany. Scientific American, December 1911.
105. Kamerlingh Onnes, H. 1911. On the change in the resistance of pure metals at very low temperatures. Communications from the Physical Laboratory of the University of Leiden **119**, 19–26.
106. Kammerlingh Onnes, H. 1915. Karol Olszewski. Chemiker Zeitung **39**, 517–524.
107. Karmann, Th. von. 1913 (Physical basis for the theory of solids, In: Encyclopedia of the mathematical sciences including their applications (*Encyclopädie der mathematischen Wissenschaften mit Einschluss ihrer Anwendungen*). Vol. 4, 45–66.
108. Karner, S. 1979. The guidance of the V2, Technikgeschichte **46**, 45–66.
109. Keesom, W. H. 1926. Prof. Dr. Kamerlingh Onnes; His Life-work, the founding of the Cryogenic Laboratory. Communications from the Physical Laboratory of the University of Leiden, Suppl. 57, 1–21.
110. Keith, S. and P. Hoch. 1982. An episode in the genesis of solid state physics. Social origins of the Bristol school of solid state theory, 1930–1939. A paper presented to the Conference on the History of Modern Physics. New Hall, Cambridge, July 1982.
111. Kevles, D. J. 1979. *The physicists; The history of a scientific community in modern America.* New York.
112. Klein, F. 1927. France and the École Polytechnique in the first decades of the Nineteenth Century. Die Naturwissenschaften, No. 1, 5–11.
113. Klemm, F. 1983. History of technology; Man and his inventions in the region of the West (*Geschichte der Technik. Der Mensch und seine Erfindungen im Bereich des Abendlandes*), Rowohlt TB 7714.
114. Comité International de Dachau. 1978. The Dachau concentration camp, 1933–1945 (*Konzentrationslager Dachau, 1933–1945*). 6th edition.
115. Kosciusko-Morizet, J. 1973. The Polytechnique "Mafia" (*La "Mafia" Polytechnicienne*). Paris.
116. Kuhn, T. S. 1967. The structure of scientific revolutions (*Die Struktur wissenschaftlicher Revolutionen*). Frankfurt.
117. Ludwig, K.-H. 1979. Technology and engineers in the Third Reich (*Technik und Ingenieure im Dritten Reich*). Düsseldorf.

118. Lux, H. 1950. Technological development and research at Telefunken during the war. **23** (No. 87/88), 11–27.
119. Manegold, K.-H. 1970. University, technical university, and industry; a contribution to the emancipation of technology in the 19th Century with special attention to the efforts of Felix Klein, (*Universität, Technische Hochshule und Industrie. Ein Beitrag zur Emanzipation der Technik im 19. Jahrhundert unter besonderer Berücksichtigung der Bestrebungen Felix Kleins*). Berlin.
120. Marx, K., and F. Engels. 1962. Works (*Werke*). Vol. 23. Berlin.
121. Mehrtens, H. 1982. The natural sciences and Prussian politics. 1806–1871. In F. Rapp and H.-W. Schutt (Eds.) Philosophy and science in Prussia; Colloquium at the Technical University of Berlin (*Philosophie und Wissenschaft in Preussen; Kolloquium an der Technischen Universität Berlin*). Pages 225–250.
122. Meyenn, K. von. 1982. Theoretical physics in the Thirties; the development of a science under ideological restrictions. Gesnerus **3/4**, 417–435.
123. Meyenn, K. von (Ed.) 1985. Wolfgang Pauli, Scientific correspondence, Volume II, 1930–39 (*Wolfgang Pauli. Wissenschaftlicher Briefwechsel, Band II, 1930–39*). Berlin.
124. Mihalovits, J. 1938. The establishment of the first teaching institution for technical mining officials in Hungary. In History of higher education in mining and forestry in Hungary, 1735–1935 (*Historia Eruditionis Superioris Rerum Metallicarum et Saltuariarum in Hungaria 1735–1935*). Sopron.
125. Möller, H. 1984. Science in emigration—quantitative and geographical aspects. Berichte zur Wissenschaftsgeschichte, **7**: 1–9.
126. Mollwo, E. 1976. Memorial address honoring Robert Wichard Pohl. In: R. W. Pohl Memorial Colloquium (*R. W. Pohl Gedächtnis-Kolloquium*). Göttingen. pp. 13–19.
127. Mollwo, E. 1981. The prehistory and early history of color center research (Zur Vor- und Frühgeschichte der Farbzentrenforschung). Unpublished manuscript. SSP.
128. Moscow Industry and Trade Newspaper, Oct. 9, 1927 (*Moskauer Industrie–u. Handelszeitung*, 9.10. 1927). German translation in SAA 68/Li 185.
129. Mott, N. F. and H. Jones, 1936. *Theory of the Properties of Metals and Alloys* (Oxford Univ. Press, 1936).
130. Mott, N. F. and R. W. Gurney, 1940. *Electronic Processes in Ionic Crystals* (Oxford Univ. Press, 1940).
131. Mott, N. F. 1961. Atomic structure and the strength of metals (*Atomare Struktur und Festigkeit der Metalle*), Braunschweig.
132. Mott, N. F. 1980. Memories of early days in solid state physics. Proc. R. Soc., London, Ser A, 371, 56–66.
133. New York Times, 1948. "The News of the Radio," p. 46.
134. OECD, 1981. The future prospects of the industrial nations (*Die Zukunftschance der Industrienationen*), OECD Report, Frankfurt.
135. Oefele, 1918. Weapons, in German natural science, technology and invention in the World War (*Deutsche Naturwissenschaft, Technik and Erfindung im Weltkriege*) Edited by Bastian Schmid, Munich.
136. Petzina, D. 1968. The policy of self-sufficiency in the Third Reich (*Autarkiepolitik im Dritten Reich*) Stuttgart.

137. Pfetsch, R. 1974. The development of science policy in Germany, 1750–1914 (*Zur Entwicklung der Wissenschaftspolitik in Deutschland 1750–1914*) Berlin.
138. Pick, H. 1976. 50 Years of color center physics, in R. W. Pohl Memorial Colloquium (*R. W. Pohl Gedächtnis-Kolloquium*) Göttingen, pp. 20–28.
139. Pick, H. 1981. Interview by J. Teichmann and G. Torkar, SSP.
140. Pirani, M. 1913. M. Pirani to W. von Siemens, 27 November 1913, SAA 4/LK 166 (Wilhelm v. Siemens).
141. Planck, Max. 1914. New avenues of knowledge, Vice Chancellor's speech, October 5, 1913. Philos. Mag. **28**, 60–71.
142. Planck, M. 1931. M. Planck to R. W. Wood, AHQP.
143. Planck, M. 1948. Scientific autobiography (*Wissenschaftliche Selbstbiographie*) Leipzig.
144. Pohl. R. W. 1913. The quantum equivalent in a photoelectric conductor. Zeitshrift für Phys. **17**, 331–346.
145. Pohl, R. W. 1928. R. W. Pohl to A. Joffe, 11 March 1928, SSP.
146. Pohl, R. W. 1933. R. W. Pohl to A. Sommerfeld, 23 November 1933 (Sommerfeld bequest, Deutsches Museum).
147. Pohl, R. W. 1936a. R. W. Pohl to R. Hilsch, 28 March 1936, SSP.
148. Pohl, R. W. 1936b. R. W. Pohl to R. Hilsch, 27 April 1936, SSP.
149. Pohl, R. W. 1936. R. W. Pohl to G. Glaser, 3 December 1936, SSP.
150. Pohl, R. W. 1938. R. W. Pohl to G. Glaser, 29 March 1938, SSP.
151. Pohl, R. W. 1938. Summary report on electron conduction and photochemical processes in alkali halide crystals. Physikalische Zeitschrift, 39, 36–54.
152. Pohl, R. W. 1939. Some recent optical investigations. Schr. Dtsch. Akademie Luftfahrtforschung, 8 1–14.
153. Pohl, R. W. 1963. Interview by T. S. Kuhn, AHQP.
154. Pohl, R. W. 1966. R. W. Pohl to H. Lorentz, 18 May 1966, SSP.
155. Prandtl, L. 1928. A conceptual model for a kinetic theory of solids. Zeitschrift für angewandte Mathematick und Mechanik 8 85–106.
156. Physical Society. 1937. Report of Conference on the Conductivity of Electricity in Solids, held at Bristol from 13th to 16th July 1937, under the joint auspices of the Physical Society and the University of Bristol. Proceedings of the Physical Society, 49, 274, Cambridge.
157. Pross, H. 1955. German academic emigration to the United States 1933–1941 (*Die Deutsche Akademische Emigration nach den Vereinigten Staaten 1933–1941*). Berlin.
158. Proszt, J. 1938. The Schemnitz Mining Academy as the birthplace of chemical and scientific research in Hungary, 1735–1935. In: History of higher education in the fields of mining and forestry in Hungary (*Historia Eruditionis Superioris Rerum Metallicarum et Saltuariarum in Hungaria, 1735–1935*). Sopron.
159. Pyenson, L. 1982. Cultural imperialism and exact sciences: German expansion overseas, 1900–1930. History of Science, **20**, 1–4.
160. Queisser, H. 1985. Crystal crises (*Kristallene Krisen*) Munich.
161. Reuter, F. 1971. Radar—its development and its use in Germany up to the end of World War II (*Funkmess—Die Entwicklung und der Einsatz des RADAR-Verfahrens in Deutschland his zum Ende des Zweiten Weltkrieges*), Opladen.

162. Ringer, F. K. 1969. *The decline of the German mandarins; the German academic community, 1890–1933*, Cambridge, Mass.
163. Robinson, A. L. 1983. Nanocomputers from organic molecules? Science **220**, 940–942.
164. Rompe, R. 1977. History of the relationships between quantum-physics and technology in the first-decades of the 20th Century. Abhandlungen der Akademie der Wissenschaften der DDR, Abteilung Mathematik–Naturwissenschaften–Technik, **7**, 13–24.
165. Rose, H., and S. Rose, 1970. *Science and society.* Middlesex.
166. Rugemer, W. 1985. New technology, old crowd—Silicon Valley (*Neue Technik, alte Gesellschaft—Silicon Valley*), Cologne.
167. Russo, A., 1981. Fundamental research at Bell Laboratories: the discovery of electron diffraction. Historical Studies in the Physical Sciences **12**, 117–160.
168. Schmid, B. 1918. German natural science, technology, and invention in the World War (*Deutsche Naturwissenschaft, Technik and Erfindung im Weltkriege*). Munich.
169. Schopman, J. 1983. Philips' response to the new semiconductor-era, germanium and silicon (1947–1957). Technikgeschichte **50**, 146–161.
170. Schottky, W. 1948. Life story (Lebenslauf). Manuscript. Schottky bequest, Deutsches Museum.
171. Schwarte, M. 1920. Technology in the World War, (*Die Technik im Weltkriege*). Berlin.
172. Science **219**, 750, 1950.
173. Scientific American, 1911; 595 and 1912; 26.
174. Seiler, K. 1948. Detectors in scientific research and medicine in Germany, 1939–1946. FIAT Reports **15**, 272–295.
175. Seitz, F. 1946. Color centers in alkali halide crystals. Reviews of Modern Physics **18**, 384.
176. Seitz, F. 1980. Biographical notes. Proceedings of the Royal Society of London, A, **371**, 84–99.
177. Sherwin, M. F. 1975. *A world destroyed; the atom bomb and the grand alliance.* New York.
178. Shockley, W. 1972. How we invented the transistor. New Scientist **56**, 689–691.
179. Siemens, W. von, 1913. W. von Siemens to Hugenberg, December 18, 1913. SAA 4/LK 166 (Wilhelm von Siemens).
180. Siemens, W. von. 1956. Recollections of my life, (*Lebenserinnerungen*), 16th edition. Munich.
181. Skarlin, H. I. 1979. *Lord Kelvin: the dynamic Victorian.* Pennsylvania, pp. 124–147.
182. Slater, J. History of the M.I.T. Physics Department 1930–1948. Unpublished manuscript, Slater bequest, American Philosophical Society Library, Philadelphia.
183. Slater, J. 1975. *Solid state and molecular theory, a scientific biography.* New York.
184. Sommerfeld, A., 1919–1960. Atomic structure and spectrum lines (*Atombau und Spektrallinien*). Braunschweig.

185. Sommerfeld, A. 1968. Collected works, edited by Fritz Sauter (*Gesammelte Schriften*, hrsg. von Fritz Sauter). 4 volumes, Braunschweig. (Cited in the text as CW I–IV.)
186. Sommerfeld bequest, Deutsches Museum.
187. Stavenhagen, L. 1985. The political boundary conditions for research (from a symposium entitled "Whither Major Research?"). Bild der Wissenschaft, 1985 No. 9, 111–129. (Reference here is to page 118.)
188. Stuewer, R. H. 1984. Nuclear physicsts in the New World; the emigrés of the 1930s in America. Berichte zur Wissenschaftsgeschichte **71**, 25–40.
189. Stuewer, R. H. 1975. *The Compton Effect, turning point in physics.* New York.
190. Struik, D. Y. 1974. Petrus von Musschenbroek, In *Dictionary of Scientific Biography,* Vol. 9, 594–597.
191. Szmborsky, K. 1984. The physics of imperfect crystals. Manuscript University of Illinois, Urbana.
192. Teichmann, J. 1983. Changing view of life (*Wandel des Weltbildes*). Munich.
193. Teichmann, J. 1985. The color center research at the First Physics Institute of Göttingen University under Robert Wichard Pohl up to 1940, its significance for ionic crystal physics and its relationship to semiconductor physics and technology (Die Farbzentrenforschung am Ersten Physikalischen Institut der Universität Göttingen unter Robert Wichard Pohl bis 1940, ihre Bedeutung im Rahmen der Ionenkristallphysik und ihre Beziehung zur Halbleiterphysik und-technik). Qualifying paper for appointment as lecturer at the University of Munich, July 1985.
194. Timoshenko, S. P. 1953. *History of strength of materials.* New York.
195. Troitzsch, U. 1976. Science and industrial practice as exemplified in the Bessemer process. In Science Report of Bielefeld University (*Wissenschaftsreport der Universität Bielefeld,* 1976, 161–175.
196. Turner, R. 1971. The growth of professorial research in Prussia, 1818–1848, causes and context. Historical Studies in the Physical Sciences **3**, 137–182.
197. Varcoe, 1970. Scientists, government and organized research in Great Britain 1914–16: The early history of DSIR. Minerva, **8**, 192–216.
198. Walker, M. The German quest for Nuclear Power 1939–59, Ph.D. dissertation, Princeton University, unpublished.
199. Weber, M. 1975. Science as a career (*Wissenschaft als Beruf*) 6th edition. Berlin. (First edition, 1919).
200. Weber, S. (Ed.) 1981. *An age of innovation; the world of electronics 1930–2000.* New York.
201. Weiner, C. 1969. A new site for the seminar: the refugees and American physics in the Thirties, In B. Bailyn and D. Fleming (Eds.) *The Intellectual Migration: Europe and America 1930–1960.* Cambridge, Mass. 1969.
202. Welker, H. 1976. Discovery and development of III-V compounds. IEEE Transactions on Electron Devices **23**, 664–674.
203. Whitney, R. W. 1921. The role of research. Scientific American, December 1921, 88–89.
204. Wien bequest. Deutsches Museum.
205. Williams, T. (Ed.) 1978. *A history of technology.* Vol. VI, Oxford.

206. Yates, R. F. 1921. Succeeding in scientific research. Scientific American, October 1921, 236.
207. Zehnder, L. 1935. W. C. Röntgen, letters to L. Zehnder (*W. C. Röntgen. Briefe an L. Zehnder*). Zurich.
208. Zeitschrift für technische Physik **1** (1920): 4f.
209. Zuckerman, S. 1966. *Scientists at war: the impact of science on military and civil affairs.* London.

SOURCES OF ILLUSTRATIONS

1. Manfred Kage photo, Institut für wissenschaftliche Fotografie, Schloss Weissenstein.
2. Copper engraving from *Histoire de l'enseignement de l'Ecole Polytechnique* (History of Education at the Ecole Polytechnique). *Livre de centenaire 1794–1894* [Centennial volume (1794–1894)], published by the Ecole Polytechnique, Vol. 1: *L'école et la science* (The school and science), Paris, 1985, p. 42.
3. Copper engraving by R. Brunet after an idea of N. le Sueur, from J. Abbé Nollet: "Essai sur l'electricité des corps" (Essay on the electricity of bodies), Paris, 1750, frontispiece.
4. Photo (1849) from the special collections in the library of the Deutsches Museum, Munich.
5. Steel engraving from A. Belloc, *La Télégraphie historique depuis les temps les plus reculés jusquà nos jours* (History of telegraphy from the most remote times to the present day), Paris, 1888, Fig. 52, p. 235.
6. Photo (1912), Siemens Museum, Munich.
7. Caricature (1908), from A. Hermann, "Weltreich der Physik; Von Galilei bis Heisenberg" (The world of physics; from Galileo to Heisenberg), *Esslingen am Nekkar,* Bechtle Verlag, 1980, plate 15 (on p. 192).
8. Photo (about 1935) from the special collections in the library of the Deutsches Museum.
9. Photo from A. Holden and P. Singer, "*Crystal and Crystal Growing,*" The Science Study Series, Vol. 6 (Heinemann Educational Books, London, 1964), plate 34.
10. Heinz Klemm photo, Pirna (German Democratic Republic).
11. Steel engraving by E. Limmer from *Illustrierte Zeitung,* Vol. 101, Leipzig-Berlin, 1893, No. 2624, p. 444.
12. Photo by C. D. Arnold (1893), Avery Architectural and Fine Arts Library, Columbia University, New York, NY.
13. Drawing by O. Gulbransson from *Simplicissimus,* Munich, 1919.
14. Atomic model from the collections of the Deutsches Museum, Munich. Subject area: Physics; branch: Atomic, Nuclear, and Elementary Particle Physics. Photo: Picture Archive, Deutsches Museum, Munich.
15. Steel engraving from L. Figuier, *Les merveilles de la science, une description populaire des inventions modernes* (The marvels of science, a popular description of modern inventions), Vol. 4, Paris, about 1870, Fig. 126, p. 217.
16. Chromolithograph from a drawing by Silbermann in a book by R. Schulze (after A. Guillemin), *Das Buch der physikalischen Erscheinungen* (The book of physical phenomena), Leipzig, 1877, plate 17 on p. 598.
17. Steel engraving from L. Graetz, *Die Elektrizität und ihre Anwendungen* (Electricity and its applications), Stuttgart, 1900, Fig. 245, 276.

18. Photo by W. Röntgen (1896), Picture Archive of the Deutsches Museum, Munich (left). Photo and marginal notes by W. C. Röntgen (Summer, 1896). Original in the Physikalisches Institut, Vienna. This reproduction taken from O. Glasser, *Wilhelm Conrad Röntgen und die Geschichte der Röntgenstrahlen* (mit dem Beitrag von M. Bovery, "Persönliches über W. C. Röntgen") [Wilhelm Conrad Röntgen and the history of X rays (with an essay M. Boveri, W. C. Röntgen the man), in the series, *Röntgenkunde in Einzeldarstellungen* (Monographs on X-radiation), edited by H. H. Berg and K. Frik], Vol. 3, Berlin, 1931, Fig. 80a, p. 242 (right).
19. Model of an air-cooling machine, Lightfoot design (1886) from the study collection of the Deutsches Museum, Munich. Photo, from the study collection in the Picture Archive of the Deutsches Museum, Munich.
20. Photo by N. Reynolds (about 1870), The Royal Institute, London.
21. Photo (1900), Kamerlingh Onnes Laboratory of the Royal University of Leyden, The Netherlands.
22. Photo (1911), Mosbach Physics Archive (E. Brüche).
23. Photo (1911), Deutsche Physikalische Gesellschaft (Picture Collection of German Physicists, Physics Archive), Bad Honnef.
24. Photo (1919) from the Sommerfeld Bequest in the special collection of the Deutsches Museum Library, Munich.
25. Atomic model produced in the shops of the Deutsches Museum, Munich, in 1925, based on data supplied by A. Sommerfeld (destroyed). Photo by the Deutsches Museum, Munich, Picture Archive.
26. Drawing from A. Sommerfeld, *Atombau und Spektrallinien* (Atomic structure and spectral lines), Braunschweig, 1919, Fig. 87, p. 367.
27. Poster from the exhibit organized by Sommerfeld's students in 1948 on the occasion of their teacher's 80th birthday at the Theoretical Physics Institute of the University of Munich. From The Sommerfeld Bequest in the special collections of the Deutsches Museum Library. Photo by the Picture Archive of the Deutsches Museum, Munich.
28. Drawings (produced by the Graphic Arts Studio of the Deutsches Museum, Munich) from A. Brachner, G. Hartl, and S. Hladky, *Atom-, Kern-, Elementarteilchenphysik. Informationen zur Ausstellung* (Atomic, nuclear, and elementary particle physics; exhibit notes), Deutsches Museum, Munich, 1983, p. 47.
29. Drawing from R. W. Pohl, *Einführung in die Mechanik und Akustik* (Introduction to mechanics and acoustics), Berlin, 1931, Fig. 46, p. 26.
30. Photo (about 1951) in the possession of Professor Mollno, Erlangen.
31. Photo (1932) from the Sommerfeld Bequest in the special collections of the Library of the Deutsches Museum, Munich.
32. Photo from R. W. Pohl, in celebration of Gauss's and Weber's electromagnetic telegraph; commemorative address delivered on November 18, 1933, at the public session of the Göttingen Scientific Society, in *Mitteilungen des Universitätsbundes Göttingen,* Göttingen, 1934, Vol. 15, No. 2, p. 7, Fig. 2.
33. (Left) Drawing from R. Hilsch and R. W. Pohl: "Control of electric currents with a three-electrode crystal and a model of a barrier layer," Z. Phys. (edited by H. Geiger in cooperation with The German Physical Society), Vol. 111, Nos. 5 and 6,

p. 407, Fig. 6, Berlin, 1938/1939. (Right) Manuscript page from the laboratory notebooks of R. Hilsch, 10/10/1938. From the Annemarie Hilsch archive, Göttingen.

34. Photo (1927) by the General Electric Company, United States . Copied from Sci. Am. **137** (No. 4), 318 (1927) (edited by O. D. Munn).
35. Advertisement of the Bell Telephone System (American Telephone and Telegraph Co. and associated companies), Sci. Am. (edited by O. D. Munn), **175** (No. 5), 195 (1946).
36. Woodcut from D. Santbech, Problematum astronomicorum et geometricorum sectiones septem...(Seven chapters on problems of astronomy and geometry...), Basel, 1561 (printed by H. Petri and P. Perna), p. 227 (below).
37. Woodcut from N. Tartaglia, Nova sciencia...(New science), Venice, 1537. Frontispiece.
38. Manuscript page of G. Galilei—1604/1609 Biblioteca Nazionale, Florence, MSS Galeleiani, Vol. 72, Folio 116v.
39. Russian aerial photograph, about 1916. Taken from Oefele, "Waffen," in *Deutsche Naturwissenschaft, Technik und Erfindung im Weltkriege* (*Weapons*, in *German Natural Science, Technology, and Invention in the World War*), edited by B. Schmid, Munich-Leipzig, 1919, p. 274.
40. Page 40 from F. P. Kerschbaum, *Die Gaskampfmittel* in *Die Technik und Erfindung im Weltkriege* (*Gas weapons*, in *Technology in the World War*), edited by M. Schwarte, Berlin, 1920, p. 278.
41. French photograph from World War I, Center for the History of Physics, American Institute of Physics (Niels Bohr Library), New York.
42. Photo (1896) from the Archives of the Roentgen Ray, London, 1897. Taken from O. Glasser, "William Conrad Roentgen und die Geschichte der Roentgenstrahlen" (mit einem Beitrag von M. Boveri, Persönliches über W. C. Röntgen) in *Röntgenkunde in Einzeldarstellungen* (William Conrad Röntgen and the history of x-rays (with an essay by M. Boveri: W. C. Röntgen the man) in Röntgenkunde in Einzeldarstellungen), monographs on x radiation, edited by H. H. Berg and K. Frik, Berlin, 1931, Vol. 3, p. 205, Fig. 63b.
43. Photo (1914–1918) from F. Dessauer, "Die Röntgentechnik im Kriege," in *Deutsche Naturwissenschaft, Technik und Erfindung im Weltkriege* (X-ray technology in the war, in German Natural Science, Technology, and Invention in the World War), edited by B. Schmid, Munich-Leipzig, 1919, plate III, following p. 794.
44. Title page from *Zeitschrift des Vereines Deutscher Ingenieure im NSBDT* (W. Parey, director), Berlin, 1944, Vol. 88, No. 1/2.
45. Advertisement of the Ethyl Corporation, New York, from Sci. Am. (edited by O. D. Munn), **168** (No. 4), 193 (1943).
46. Title page from Sci. Am. (edited by O. D. Munn), **168** (No. 2), (1943).
47. Photo from *Elektrotechnische Zeitschrift, Organ des Verbandes Deutscher Elektroniker* (VDE-Verlag, Wuppertal, 1977), Vol. 29, No. 24, p. 779.
48. Photo from *A History of Engineering and Science in the Bell System: National Service in War and Peace 1925–1975*, edited by M. D. Fagen (Bell Telephone Laboratories, Murray Hill, NJ, 1978), p. 45 (below), Fig. 2-19.

49. Advertisement of the Bell Telephone System (American Telephone and Telegraph Co. and associated companies), Sci. Am. (edited by O. D. Munn), **166** (No. 4), 165 (1942).
50. Advertisement of the Bell Telephone System (American Telephone and Telegraph Co. and associated companies), from Sci. Am. (edited by O. D.Munn), **168** (No. 4) 149 (1943).
51. Photo from *A History of Engineering and Science in the Bell System; National Service in War and Peace 1925–1975,* edited by M. D. Fagen (Bell Telephone Laboratories, Murray Hill, NJ, 1978), p. 213, Fig. 4-19.
52. Photo (about 1943/44) from *A History of Engineering and Science in The Bell System; National Service in War and Peace 1925–1975,* edited by M. D. Fagen (Bell Telephone Laboratories, Murray Hill, NJ, 1978), p. 110, Fig. 2-56.
53. Plant photo (1948), Western Electric Co., United States.
54. Schematic drawing of the first point-contact transistor (after J. Bardeen and W. Brattain—1947), from L. Hoddeson, "The Discovery of the Point-Contact Transistor," in *Historical Studies in the Physical Sciences,* edited by J. L. Heilbron (University of California Press, Berkeley, 1981), Vol. 12, No. 1, p. 74, Fig. 8.
55. Photo (1947), Bell Telephone Laboratories (United States).
56. Advertisement of the Bell Telephone Laboratories, in Sci. Am. (edited by D. Flanagan), **189** (No. 1), 9 (1953).
57. Advertisement for RCA (the Radio Corporation of America) in Sci. Am. (edited by D. Flanagan), **185** (No. 4) (1951).
58. Photo from "Project Tinkertoy," *Funktechnik, Fernsehen, Electronik,* Berlin, 1954, Vol. 9, p. 24.
59. Advertisement of Tung-Sol Electric, Inc., in Sci. Am. (edited by D. Flanagan), **204** (No. 1), 167 (1961).
60. Advertisement of General Electric Co., in Sci. Am. (edited by D. Flanagan), **204** (No. 3), 167 (1961).
61. Advertisement of Leeds and Northrup, in Sci. Am. (edited by D. Flanagan), **204** (No. 1), 209 (1961).
62. Advertisement of Transitron Electric Corporation, in Sci. Am. (edited by D. Flanagan), **204** (No. 1), 85 (1961).
63. Manuscript pages from the notebooks of J. Kilby (1958/59). Taken from "Es war einmal...Die Geschichte des Mikroprozessors" (Once upon a time...The History of the microprocessor." CHIP No. 11, 20 (excerpt) (1981), Würzberg, Vogel-Verlag.
64. Plant photo, Texas Instruments (United States).
65. Photo by O. P. Herrnkind from his article, "Hivac Miniaturröhren" (Hivac miniature tubes), in Funk Ton (edited by G. Leithauser), **5**, 218 (1948).
66. Photo (about 1948), Bell Telephone Laboratories (United States).
67. Photo (about 1963), Mullard Limited (United States).
68. Advertisement of the Philips firm, in Sci. Am. (edited by D. Flanagan), **246** (No. 2), 4 (1982).
69. Photo (1946), Siemens Museum, Munich.
70. Photo (1977), Ben Rose (United States). From R. N. Noyce, Sci. Am. (edited by D. Flanagan), **237** (No. 3), 66 (1977).

71. Photo (1945), Western Electric Company (United States).
72. Plant photo, Siemens AG, Villach, Austria.
73. Photo (1980), C. O'Rear/West Light/Focus, Hamburg.
74. Drawing based on sketch by the authors.
75. Drawing based on sketch by the authors.
76. Drawing based on sketch by the authors.
77. Drawing from A. Sommerfeld and H. Bethe, "Elektronentheorie der Metalle," in *Handbuch der Physik* (Handbook of Physics), edited by H. Geiger and K. Scheel, Berlin, 1933, Vol. 24/2, Chap. 3, 401, Fig. 25c.
78. Drawing from C. Kittel, *Einführung in die Festkörperphysik* (Introduction to solid state physics), Vienna, R. Oldenbourgh-Verlag, 1968, Chap. 9, p. 353 (left), Fig. 32. Drawing by M. Puebla after L. M. Falicov, Phys. Rev. Lett. 7, 231 (right) (1961).
79. Drawings based on H. Stroppe, *Physik für Studenten der Natur- und Technikwissenschaften* (Physics for students of the natural and engineering sciences), 1984, p. 426, Fig. 6.42(a,b,c).
80. Drawing from K. W. Dregge and D. Haferkamp, *Grundlagen der Electronik* (Fundamentals of electronics) (Vogel Buchverlag, Würzburg, 1985), p.97.
81. Drawing from H.-U. Harten, N. Nägerl, and H.-D. Schulte, *Festkörperphysik (Solid state physics)* (Verlag Herder, Vienna, 1978), p. 106.

TABLES

I. Drawing based on F. R. Pfetsch, *Zur Entwicklung der Wissenchaftspolitik in Deutschland 1750–1914* (The development of science policy in Germany, 1750–1914) (Duncker & Humblot, Berlin, 1974), Chap. 6, p. 273, Fig. 1.

II. Drawing based on F. R. Pfetsch, *Zur Entwicklung der Wissenschaftspolitik in Deutschland 1750–1914* (The development of science policy in Germany, 1750–1914), (Duncker & Humblot, Berlin, 1974), Chap. 6, p. 282, Fig. 3.

III. Drawing from H. Kamerlingh Onnes, Univ. Leiden No. 124, 23 (1911).

IV. Drawing from K. H. Hellwege, *Einführung in die Physik der Atome (3. Verbesserte Auflage)* [Introduction to atomic physics (Third improved edition)] (Springer-Verlag, Berlin, 1970), Vol. 2, p. 81, Fig. 35.

V. Based on A. S. Beyerchen, *Wissenschaftler unter Hitler; Physiker im Dritten Reich* (Scientists under Hitler; physicists in The Third Reich) (Verlag Kiepenheuer und Witsch, Cologne, 1980), pp. 64 and 65.

VI. From S. R. Weart, "The Physics Business in America, 1919–1940: A Statistical Reconnaissance, in *The Sciences in the American Context: New Perspectives,* edited by N. Reingold (Smithsonian Institute Press, Washington, DC, 1979), p. 296, Fig. 1.

VII. From *A History of Engineering and Science in the Bell System; National Service in War and Peace 1925–1975,* edited by M. D. Fagen (Bell Telephone Laboratories, Murray Hill, NJ, 1978), p. 11, Fig. 1-2.

VIII. Based on J. E. Tilton, *International Diffusion of Technology: The Case of Semiconductors* (Brookings Institution, Washington, 1971), p. 66. From E. Braun and S. MacDonald, "Von der Erfindung zur Innovation—dargestellt am Beispiel Halbleiter—Elektrotechnik (T.2—Die Innovation)" [From invention to innovation—as exemplified in the case of semiconductor engineering (Part 2—Innovation)], in *Elektrotechnische Zeitschrift—B; Fachzeitschrift für Anwendung und Betrieb,* edited by the Verband Deutscher Elektrotechniker (VDE-Verlag, Berlin, 1977), Vol. 29, No. 26, p. 856, Table I.

IX. Based on data in Don C. Hoefler, Semiconductor equipment and materials, Inc., Mountain View (California), 1979. Here based on E. Braun and S. MacDonald, *Revolution in miniature; The History and Impact of Semiconductor Electronics Re-explored in an Updated and Revised Second Edition* (Cambridge University Press, Cambridge, 1982). p. 127, plate 10.4.

X. Based on data in J. Tilton, *International Diffusion of Technology: The Case of Semiconductors* (Brookings Institution, Washington, 1971); *Market Data Book,* (Electronic Industries Association, Washington, 1975). Here based on E. Braun and S. MacDonald, *Revolution in Miniature: The History and Impact of Semiconductor Electronics Re-explored in an Updated and Revised Second Edition* (Cambridge University Press, Cambridge, 1982), p. 98, plate 8.2.

XI. From R. Curow and S. Curran, " Anwendung der Technologie" in *Auf Gedeih und Verderb; Mikroelektronik und Gesellschaft; Bericht an den Club of Rome* (Application of technology in For better or for worse; Microelectronics and society; Report to the Club of Rome), edited by G. Friedrichs and A. Schaff (Rowohlt Taschenbuch-Verlag GmbH, Reinbek bei Hamburg, 1984), pp. 128 and 129.

XII. Based on H. C. de Mattos, *Technology and developing countries in the International Telecommunications Union*, 3rd World Telecommunication Forum, Geneva, 1979 (Rowohlt, Reinbek, 1984), Vol. 1, p. 11. Here from J. Evans, "Arbeitnehmer am Arbeitsplatz" in *Auf Gedeih und Verderb; Mikroelektronik und Gesellschaft; Bericht an den Club of Rome* (Employees in the workplace, in For better or for worse; Microelectronics and society; Report to the Club of Rome), edited by G. Friedrichs and A. Schaff (Rowohlt Taschenbuch-Verlag GmbH, Reinbek bei Hamburg, 1984), p. 185.

XIII. Drawing from Science **219**, 750 (1983) (American Association for the Advancement of Science, Washington, DC, 1983.

XIV. Drawing from M. Bianchi-Streit, N. Blackburne, R. Budde, H. Reitz, B. Sagnell, H. Schmied, and B. Schorr, *Der wirtschaftliche Nutzen der CERN—Aufträge an die Industrie (Zweite Studie)* [The economic benefit of CERN contracts with industry (second study)], Geneva, CERN (European Organization for Nuclear Research), Scientific Information Service, 1985, pp. 18 and 19 (top).

XV. Drawing from M. Bianchi-Streit, N. Blackburne, R. Budde, H. Reitz, B. Sagnell, H. Schmied, and B. Schorr, *Der wirtschaftliche Nutzen der CERN—Aufträge an die Industrie (Zweite Studie)* [The economic benefit of CERN—contracts with industry (second study)] Geneva, CERN (European Organization for Nuclear Research), Scientific Information Service, 1985, p. 19.

INDEX